THE GRAND TOUR

Saturn from Tethys

THE GRAND TOUR
A·TRAVELER'S·GUIDE·TO·THE·SOLAR·SYSTEM

By Ron Miller & William K. Hartmann

WORKMAN PUBLISHING · NEW YORK

Library of Congress Cataloging in Publication Data
Miller, Ron, 1947-
   The grand tour.

   Bibliography: p.
   Includes index.
   1. Solar system. I. Hartmann,
      William K.
II. Title.
QB501.2.M54   523.2   80-54620
ISBN 0-89480-147-3   AACR2
ISBN 0-89480-146-5 (pbk.)

Manufactured in the United States of America
First printing September 1981
10 9 8 7 6 5 4 3 2 1

Workman Publishing Company, Inc.
1 West 39 Street
New York, New York 10018

Acknowledgments

The authors are indebted to many individuals for their help. We thank Jay Inge and R. A. Batson of the U.S. Geological Survey for providing the shaded relief maps of the planets and their moons; Les Gaver, NASA Headquarters, for the photographs taken by unmanned probes and *Apollo* and *Skylab* astronauts; Chris Spielmann of Advanced Images, Inc., and Brian Sullivan and Dennis Mammana of the Flandrau Planetarium staff for assistance in photography of the artwork; Judith A. Miller, who critiqued and typed the manuscript; Gayle and Amy Hartmann, Pamela Lee, Floyd Herbert, and the staff of the Planetary Science Institute of Science Applications, Inc., for critiques and suggestions; Louise Gikow, who bravely edited the text; and Ian Summers, Peter Workman, and Sally Kovalchick for faith, enthusiasm, support, and unlimited patience.

Cover and book design by T. O. Miller
Cover illustration by Ron Miller

Dedication

Without Chesley Bonestell (1888-∞), this work, in so many ways, would not have been possible; and it is to him, with humble appreciation, that this book is dedicated. (Whether he likes it or not.)

CONTENTS

CONTENTS

INTRODUCING THE TERRITORY

A NEW LOOK AT THE SOLAR SYSTEM

Far out in interstellar space, our ship flies among millions of stars strewn along one curving arm of a spiral galaxy. We are searching for planetary systems. A *planet* is a nonluminous body that circles a star. Very small planets, such as asteroids, are called *minor planets*. Moons, or *satellites* (which means "companions") orbit around planets, which in turn orbit stars.

We pick an average star. The star is our sun. Our goal is to see our own planetary system as a new visitor might perceive it, not centered around Earth, but with Earth as only one planet out of many. Like naturalists of old, like Charles Darwin voyaging indiscriminately from island to island, we want to see what all these places are like. What grandeur, desolation, power, silence, resources, or loneliness does each offer?

We scrutinize the approaching star. At first we can see no planets at all, because the sun is vastly brighter and more massive than any of them. The planets are lost in the sun's glare.

We try our radio. It picks up faint signals at radio, TV, and other wavelengths. These signals might not necessarily reveal intelligent life in the system: they might be network TV broadcasts, or the random snaps, crackles, and pops of electrical activity like lightning in the atmospheres of Jupiter, Venus, and Earth. But they at least suggest the presence of planets.

We can also search for planets at this distance by looking for their gravitational effects. As they orbit around the sun, the planets tug it slightly to and fro. Sensitive instruments can pick up this "wiggle" in the sun's position. The major wiggle is due to the largest planet, massive Jupiter. Jupiter has more mass, hence gravity, than all the other planets put together. So at first we only notice Jupiter's effect, and conclude that the sun is essentially a double system. The largest object in our system—the sun—is said to have 1 solar mass; the second largest object— Jupiter—has about 0.001 solar mass. The rest of the planets put together have only 0.0004 solar mass—easy to overlook from our perch out in space.

With both radio and gravity data indicating planets in this system, we come in for a closer look. When we first catch sight of a planet, it is again Jupiter, five times as far from the glaring sun as Earth is. Astronomers, following the old Earth-chauvinist tradition of our forebears, defined the standard unit of distance in the solar system as the distance from the sun to Earth, 150 million kilometers or 93 million miles. This is called one *astronomical unit*. So we say that Earth is 1 AU from the sun; Jupiter is 5.2 AU from the sun.

Most stars have companion stars orbiting around them. As we catch sight of Jupiter, we wonder whether this object should be classified as a planet or a small companion star to the sun. What is the difference? Stars shine by means of nuclear reactions inside of them. Planets are not big enough to generate the heat and pressure necessary to cause nuclear reactions in their centers, so planets don't shine. You might think it is obvious that Jupiter is a planet, but telescopic observers discovered in the 1960s that Jupiter radiates several times more energy than it receives from the sun. This means that energy *is* being created somehow inside Jupiter, and in this sense Jupiter is like a star. But theoretical work shows that the energy is not coming from nuclear reactions. Instead, it is being generated by a slow contraction that has been going on since Jupiter formed. Jupiter's gravity is so strong that it keeps pulling itself inward, liberating a modest amount of heat at the same time.

This theoretical work also indicates that a body must be about eighty times as massive as Jupiter in order to generate enough central heat and pressure to initiate nuclear reactions and become a true star. Thus, we can say that the solar system has only one star. Jupiter has quite a long way to go before it could accumulate enough material to become a star. If more had been available to proto-Jupiter, the solar system might have had two suns, making our astronomical climate strikingly different.

NINE PLANETS?

If you study the sky from night to night, you will find that the planets—such as diamond-bright Venus or ruddy Mars—move among the stars. This is why the Greeks gave them the name *planētēs*, or wanderers. They thought that the planets moved around Earth in complicated loops. In 1543, the Polish astronomer Nicolaus Copernicus showed that the planets' movements could be better understood by postulating that they moved around the sun. Acceptance of this idea came to be called the Copernican revolution.

It was an epoch-making breakthrough, because it affected how we think about ourselves in relation to the universe. For the first time, we were forced to face the idea that the universe doesn't revolve around us. We are just a part of it.

Many people seem to assume that the Copernican revolution was the one and only revolution in our view of the solar system, and that things have been stable since then. More planets have been discovered in more or less circular orbits around the sun, of course. And more moons have been found in orbits around planets. But for years, textbooks have reported that there are nine planets, and most people have the impression that the hierarchy of planets, moons, and interplanetary debris is very clear-cut, with the planets being the big, important objects, the moons lesser, and so on.

Today, we are beginning to look at the solar system in a new way, a break from the "nine planets gestalt." Our recent spacecraft voyages have made us realize that the solar system has many more than nine large worlds. Some of the moons in it are bigger than some of the

planets. There are about twenty-five worlds larger than a thousand kilometers, or 620 miles, across. We will visit those worlds. Numerous interplanetary bodies, neither planets nor moons, range up to a thousand kilometers across. We will visit the most interesting of these smaller worlds. These are traditionally classified as *asteroids, comets,* and *meteoroids.* According to the traditional observational distinctions, asteroids are stony bodies one to a thousand kilometers in size, situated mostly between Mars and Jupiter. Comets, one to a hundred kilometers in size, are icy bodies that are mostly located in the outermost solar system, but sometimes drop into the inner solar system, where the sun melts off some of their ices and releases the gas and dust that streams out to form a cometary tail. Meteoroids are fragments of asteroids and comets, typically microscopic to a hundred meters in size. Meteoroids that strike planetary surfaces are called *meteorites* and are composed of various types of stone or nickel-iron metal.

Interplanetary bodies are too numerous to count. Catalogs of objects with well-defined orbits include over two thousand asteroids and several hundred comets. The solar system contains thousands of bodies big enough to land on, though the gravity of the small ones is so

weak that you might wish to tether your spacecraft so as not to have it drift off. And you should be careful not to launch *yourself* off such a body by an overenthusiastic jump.

Data from the 1970s and early '80s makes us realize that the traditional practice of categorizing each object as a planet, moon, asteroid, comet, or meteoroid is misleading. It blurs their relationships by setting up different pigeonholes, obscuring the unique nature of each. Spectral observations, for instance, show a wide range of different rock types among different asteroids. Moons and planets vary in density, composition, and surface type. Some moons seem more closely related to some asteroids than to other moons. Scientific arguments have been fought over whether certain "asteroids" are really asteroids at all, or simply burnt-out comets. Other research focuses on whether meteorites are pieces of comets, asteroids, or a mixture of fragments from both sources.

If we reject the category-prone way of looking at things, we can see the variety and relationships of planetary bodies more clearly.

A COMPOSITIONAL SEQUENCE

One relationship we find among major planetary bodies is a sequence of compositions that varies primarily with their

different distances from the sun. The planets were formed by an aggregation of dust grains condensed out of cooling gases that surrounded the newly formed sun. These gases were hottest near the sun and coolest away from the sun.

Mineral grains requiring low temperatures to condense could not do so near the sun; those that condense at higher temperatures were the only ones to solidify in the inner solar system. These included metal grains and particles of various silicate minerals like those found in Earth rocks, lunar rocks, and meteorites. Farther from the sun, such grains were accompanied by lower-temperature compounds, such as carbon-rich minerals and hydrated minerals. Beyond the asteroid belt, the gas was cold enough for ice (frozen water) to form. Out beyond the vicinity of Uranus or Neptune, even gases such as methane and ammonia froze, forming ice grains. The inner solar system worlds (out to the asteroid belt) are therefore mostly rock with some metals, while the outer solar system worlds (beyond the belt) are rich in ices of various kinds.

Now we can see a new way to perceive the asteroid/comet distinction. Asteroids are the stone objects that originally formed in the inner solar system. Comets are the more ice-rich bodies that originally formed in the outer solar system.

Gravitational perturbations by the massive planets have thrown many of these small bodies into parts of the solar system far from where they first formed. Thus, some stony stragglers may be found today in the outer solar system, while icy comets drop into Earth's vicinity for an occasional visit.

Current data suggest that large satellite systems, like Jupiter's, may have formed like miniature solar systems, and also may share the same compositional patterns. From the measured densities of Jupiter's large moons, we know that the inner ones have mostly rocky interiors, while the outer ones have icy interiors. Jupiter's radiation was probably a significant heat source during the formation of satellites around it, just as the sun heated the inner solar system.

When an inner planet experiences internal melting (generally due to radioactivity heating its interior), it produces silicate lavas like those of Earth and its moon. But when an outer world melts, the erupted "lava" may well be water, and the "frozen lava flows," sheets of ice. This probably explains why many worlds in the outer solar system have icy surfaces. In particular, some of the inner moons of giant planets, though they may have stone-rich interiors, seem to have been heated enough to produce watery eruptions that left ice on their surfaces.

Evidence from meteorites and the moon shows that planets aggregated from cool, solid material 4.5 billion years ago. Evidence from lunar rocks indicates that they had molten layers at their surfaces almost as soon as they formed, but that these so-called magma oceans cooled quickly, by about 4.4 or 4.3 billion years ago. All rocky planetary material contained small amounts of radioactive minerals, which produced small amounts of heat. The smallest planets could radiate this heat easily and cool quickly. Therefore, planetary bodies smaller than a few hundred kilometers across probably never either melted, or developed volcanoes; their surfaces are much the same as they were when they were formed, still preserving the impact scars of the meteorites that fell on them and helped mold them.

The larger worlds could not shed their heat so easily, because planetary bulk insulated their hot centers. These centers got hotter and hotter. The insides of worlds larger than 1,000 to 2,000 kilometers (620 to 1,240 miles) melted around 4 billion years ago, allowing heavy molten iron to sink into them, forming metal cores. The light "slag" of silicates and watery material floated to the top. These planets thus gained crusts of silicate-rich rocks that tend to have low densities and light coloration.

Volcanism broke through these crusts, covering parts of the surfaces with lava flows. On the smaller worlds, interiors soon cooled and volcanism declined. Volcanism continued for a longer period of time on larger worlds.

Often, the last eruptions were from deeper in the planet and produced rocks poorer in silicates and darker in color than the crustal rocks. On the moon (diameter 3,476 kilometers, or 2,155 miles), such eruptions formed dark lava plains covering about 15 percent of its surface. They ended about 3 billion years ago, as the moon's interior cooled. On Mars (diameter 6,795 kilometers, or 4,212 miles), internal heat lasted longer, promoting volcanic eruptions as recently as 1 billion years ago or less. These eruptions resurfaced about half the planet. On Earth (12,756 kilometers, or 7,908 miles), the eruptions have continued through geologic time. The primeval crust has been totally destroyed, and volcanoes continue to erupt.

In addition to internal evolution, atmospheric evolution also depends on the size of the planet. Small planets have too little gravity to retain atmospheres. Gases released by small planets' volcanoes escape into space. Planets larger than about 6,000 kilometers (3,720 miles) have enough gravity to hold down the heavier volcanic gases, such as carbon dioxide. Mars, for instance, has a thin atmosphere of carbon dioxide; some of the molecules escape into space, one at a time. Venus has a thick carbon dioxide atmosphere. Earth has a more complex atmosphere in which much of the carbon dioxide has been broken down by plants, thus releasing oxygen. Light gases, such as hydrogen, have mostly risen to the top of our atmosphere and leaked into space, while only heavier gases have been retained. Giant Jupiter, 143,000 kilometers (88,660 miles) across (eleven times bigger than Earth) has a deep, dense atmosphere. Even hydrogen, the most abundant but lightest element in the solar system, is retained by Jupiter. Because the larger planets have denser atmospheres, their surfaces are more modified by erosion. Earth has lost virtually all of its meteorite impact craters to erosion and tectonics; Mars has dust dunes and dry streambeds. Only airless bodies like the moon preserve surfaces saturated with craters formed during the early days of the solar system.

A PLAN FOR OUR TOUR

If we organized our tour of the solar system according to tradition, we might start close to the sun and work our way outward. But then we would go from moonlike Mercury to cloud-shrouded Venus; from tiny asteroids to massive Jupiter. The scenery would be a chaotic hodgepodge, because scenery is controlled by the evolutionary state of a world, and evolutionary state is controlled largely by size.

So we propose to take a new kind of tour: a tour in order of size. First comes massive Jupiter, the sun's principal companion. Next is the second most massive planet, Saturn; and so on. This approach allows us to progress from energetic, massive, active, evolved worlds to dormant, small, primitive worlds with empty craters that seem to echo with the explosions of meteorite impacts billions of years old.

There is a second advantage of touring our solar system in this order. It will soon become clear that our planetary neighborhood does *not* consist of the traditional nine planets and insignificant, smaller worlds. We will visit all twenty-five worlds larger than a thousand kilometers across, of which eighteen are planets or moons bigger than the planet Pluto.

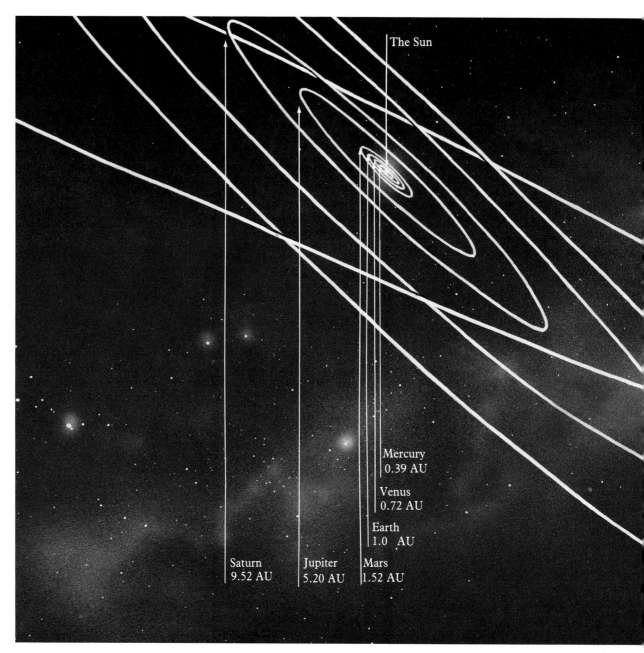

The Sun

Mercury
0.39 AU

Venus
0.72 AU

Earth
1.0 AU

Mars
1.52 AU

Jupiter
5.20 AU

Saturn
9.52 AU

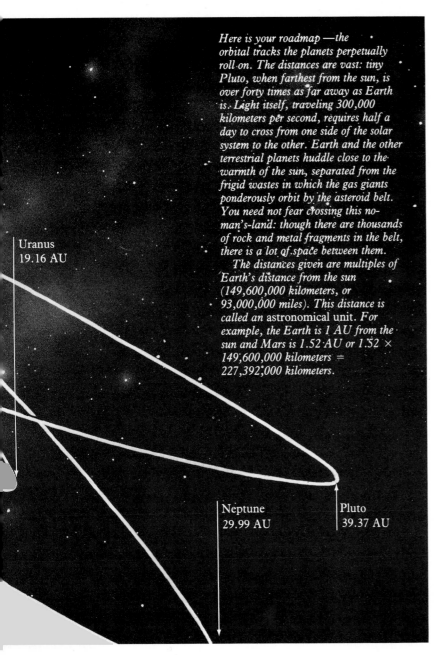

Here is your roadmap—the orbital tracks the planets perpetually roll on. The distances are vast: tiny Pluto, when farthest from the sun, is over forty times as far away as Earth is. Light itself, traveling 300,000 kilometers per second, requires half a day to cross from one side of the solar system to the other. Earth and the other terrestrial planets huddle close to the warmth of the sun, separated from the frigid wastes in which the gas giants ponderously orbit by the asteroid belt. You need not fear crossing this no-man's-land: though there are thousands of rock and metal fragments in the belt, there is a lot of space between them.

The distances given are multiples of Earth's distance from the sun (149,600,000 kilometers, or 93,000,000 miles). This distance is called an astronomical unit. For example, the Earth is 1 AU from the sun and Mars is 1.52 AU or 1.52 × 149,600,000 kilometers = 227,392,000 kilometers.

Uranus
19.16 AU

Neptune
29.99 AU

Pluto
39.37 AU

WHAT TO LOOK FOR

We are about to embark on a tour of worlds—a tour that will enable us to change our ways of thinking about planets. We are going to try to see each world on its own terms, one at a time.

What we are really looking for is a new perspective on our solar system, and a new sense of the fantastic variety of wilderness areas that surround us, waiting just over the horizon. . . .

Pick your home planet out of this celestial lineup of giants and dwarves (overleaf). A hint: it doesn't look as significant as you might think it should. The enormous golden landscape across the bottom is all of the sun that will fit into the picture at the same scale as the planets. The pretty blue planet third from the left would drop into one of the whirling sunspots like a billiard ball into a pocket.
Two distinct planetary groups immediately stand out. One consists of four small planets—Mercury, Venus, Earth, and Mars—that share rocky mantles covering metallic cores. Since they more or less resemble Earth, they are called the terrestrial planets. *The other group is hard to miss: the gas giants—Jupiter, Saturn, Uranus, and Neptune. These four planets contain over two hundred times the mass of all the rest of the planets combined. Pluto doesn't seem to fit into either group; it may be a moon Neptune lost long ago. Lined up with their parents like proud children are the more than thirty-seven moons of the solar system. Some of these are so large that if they orbited the sun rather than a planet, they could be considered planets in their own right.*

Below is a key to the numbered satellites on the painting overleaf. For a complete listing of all the planets and satellites in the solar system, as well as several major asteroids, see the charts at the back of the book.

Earth
1. The Moon

Mars
1. Phobos
2. Deimos

Jupiter
1. 1979 J1   8. Himalia
2. Amalthea   9. Lysithea
3. Io   10. Elara
4. Europa   11. Anake
5. Ganymede   12. Carme
6. Callisto   13. Pasiphae
7. Leda   14. Sinope

Saturn
1. Numerous small satellites
2. Mimas   7. Titan
3. Enceladus   8. Hyperion
4. Tethys   9. Iapetus
5. Dione   10. Phoebe
6. Rhea

Uranus
1. Miranda   4. Titania
2. Ariel   5. Oberon
3. Umbriel

Neptune
1. Triton
2. Nereid

Pluto
1. Charon

MERCURY   VENUS   EARTH   MARS   JUPITER

1

1
2

14 13 12 11 10 9 8 7 6 5 4 3 2 1

THE SUN

Ron Miller

A Family Portrait

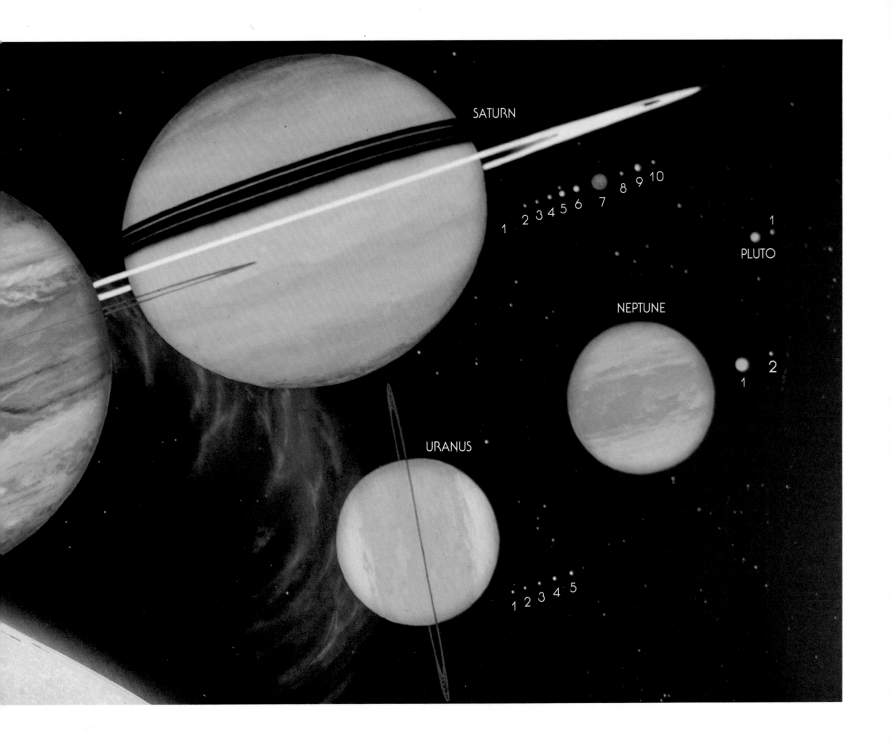

SATURN

1 2 3 4 5 6 7 8 9 10

PLUTO
1

NEPTUNE
1 2

URANUS
1 2 3 4 5

JUPITER PLANET OF THE GODS

Average distance from the sun:
778,300,000 km.
Length of year: 11.86 years
Length of day: 9 hours 55 minutes
Diameter: 143,200 km.
Surface gravity (Earth = 1): 2.3
Composition: hydrogen; ices; hydrogen
atmosphere

Jupiter was the giant among gods, and it is also the giant among planets. It has about two and a half times the mass of all other planets combined, and therefore has the strongest gravitational pull, affecting the motions of many other planets and interplanetary bodies. The second largest planet, Saturn, has only 84 percent the diameter of Jupiter and less than one-third its mass. Earth would comfortably nest inside the largest oval storm system that swirls in Jupiter's skies.

Jupiter is covered with dense clouds in a thick atmosphere estimated to contain about 88 percent molecular hydrogen gas and 11 percent helium gas. Minor constituents include methane, ammonia, water, carbon monoxide, and other compounds. Interestingly, such an atmosphere is probably similar in composition to that which enveloped Earth in the first 100 million years or so of its history. (This may sound like a long time, but is only 2 percent of the age of the planets.) The planets were originally formed from a nebula made up of these gases; Jupiter is so big that its strong gravity has trapped and held them, while on Earth and smaller bodies the fast-moving hydrogen molecules have randomly zipped off into space, one at a time. Earth's atmosphere has thus thinned and evolved in composition; Jupiter's has remained in nearly its original state.

Many of Jupiter's hydrogen atoms, which are chemically active, have combined with less abundant elements to form exotic, brightly-colored compounds. These lend their vivid hues to the clouds, which are composed of ammonia "snowflakes," water "snowflakes," and "snowflakes" of compounds such as ammonium hydrosulfide. Red and orange clouds that formed at certain altitudes and latitudes swirl into white and tan clouds formed in other regions. Jupiter rotates more quickly than Earth, in about ten hours, and its rotational dynamics shear the cloud patterns out into long bands and zones that run around the planet parallel to its equator. The two most prominent dark bands of clouds are the North and South Equatorial Belts, a few degrees on either side of a lighter Equatorial Zone, and additional dark belts and light zones also ring the planet.

When we plunge into Jupiter's atmosphere, the sky turns lighter and bluer as we approach the cloud tops because of scattered sunlight in the air—the same process that makes the sky blue on Earth. We begin to encounter high clouds at a level where the pressure is almost equal to that on Earth's surface, but the temperature is only around − 123 degrees Centigrade (C), or − 189 degrees Fahrenheit (F). As we descend into the clouds, the air grows warmer and denser, reaching room temperature among cloud layers with an air pressure perhaps five times greater than that on Earth. As we keep going downward, what do we find?

Jupiter's atmosphere is far too dense to reveal its surface to telescopes. Theoretical studies of Jupiter's chemistry and physics suggest that the atmosphere simply gets thicker and thicker the closer you get to the planet's surface. It may grade eventually into a vast dark fog of water-ammonia droplets, overlying a sluggish ocean of liquid hydrogen and perhaps other compounds.

Atmosphere dynamics and surface conditions depend partly on the amount of heat flowing through an atmosphere. On Earth, this heat is almost entirely associated with sunlight coming from above and reradiating off the ground. In the 1960s, astronomers were surprised to discover that Jupiter radiates several times more energy than it absorbs from sunlight. The extra energy must be coming from Jupiter's interior. Further theoretical studies showed that

Ron Miller

Jupiter's ring (above) surrounds Jupiter like an enormous smoke ring, visible only when the sun is shining directly behind it. With the exception of some small rubble, Jupiter's ring is composed of particles no bigger than those in *cigarette smoke—literally microscopic dust. The ring is material slowly spiraling into Jupiter, but may be constantly being replenished by the erosion of a small moon that sits on the ring's outer edge. The distant sun (only* *one-fifth its diameter as seen from Earth) illuminates Jupiter and two of its large moons in narrow crescent phases.*

15

JUPITER

Jupiter's ocean (right) *is a sluggish, oily mass of liquid hydrogen, barely able to raise a few slow, rolling waves against the almost unimaginable pressure of hundreds of kilometers of Jupiter's atmosphere. The frigid, tideless ocean covers the entire planet; it is a world sea 114 times the area of the entire planet Earth. A cruise ship making twenty knots would take only one and a half months to circumnavigate Earth, but nearly one and a half years to circle Jupiter. Little or no sunlight filters down through the dense blankets of clouds, but mighty blasts of lightning in the lower clouds occasionally illuminate the Stygian, endless sea.*

Ron Miller

16

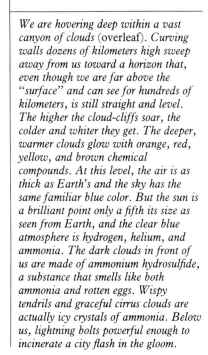

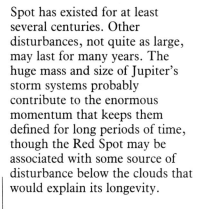

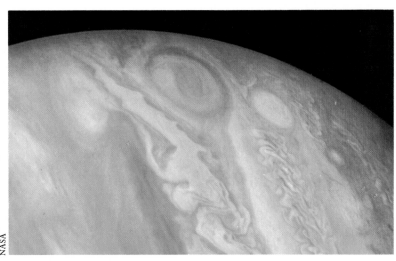

Jupiter's Great Red Spot (above) *is an enormous hurricane large enough to engulf Earth.*

this energy release probably involves a slow contraction of the planet—a long-term process that began as Jupiter formed, creating heat by compressing the inside of the planet. Some of the heat that was initially produced is still leaking out, and new heat is constantly being generated.

This extra energy contributes a certain amount of warmth from below Jupiter's clouds, and increases activity in the cloud layer. If you heat a pan of water or cooking oil on the stove, you will note that the material begins to circulate long before any boiling starts. This thermally created movement is called a *convection current.* Similarly, convection due to the heat from

Jupiter's surface stirs giant, rapidly moving masses of gas in its atmosphere. This may bring colored material up from below, initiate condensation of new cloud particles at higher levels, and lead to cloud patterns that swirl around in oval hurricanes or get sheared into long, convoluted strips along the belts and zones.

The most famous such disturbance is the Great Red Spot, a brick-red cloud system larger in diameter than Earth. Spacecraft observations have shown a vigorous circulation in the Red Spot. Smaller clouds nearby get sucked into the Red Spot, circulate in it over a period of hours, and finally get torn apart. Other clouds are almost sucked in, circulate around the periphery, and are then spit out into adjacent zones. The Red

Spot has existed for at least several centuries. Other disturbances, not quite as large, may last for many years. The huge mass and size of Jupiter's storm systems probably contribute to the enormous momentum that keeps them defined for long periods of time, though the Red Spot may be associated with some source of disturbance below the clouds that would explain its longevity.

We are hovering deep within a vast canyon of clouds (overleaf). *Curving walls dozens of kilometers high sweep away from us toward a horizon that, even though we are far above the "surface" and can see for hundreds of kilometers, is still straight and level. The higher the cloud-cliffs soar, the colder and whiter they get. The deeper, warmer clouds glow with orange, red, yellow, and brown chemical compounds. At this level, the air is as thick as Earth's and the sky has the same familiar blue color. But the sun is a brilliant point only a fifth its size as seen from Earth, and the clear blue atmosphere is hydrogen, helium, and ammonia. The dark clouds in front of us are made of ammonium hydrosulfide, a substance that smells like both ammonia and rotten eggs. Wispy tendrils and graceful cirrus clouds are actually icy crystals of ammonia. Below us, lightning bolts powerful enough to incinerate a city flash in the gloom.*

Ron Miller

Jupiter's Clouds

Ron Miller

20

Jupiter probably has a central core of rocky material resembling an oversized Earth, overlaid by an enormous mantle of a form of high pressure liquid hydrogen called *liquid metallic hydrogen*, which can conduct electric current and may be involved in producing Jupiter's very strong magnetic field. Covering this is an ocean of liquid hydrogen in an ordinary molecular state. Whether the surface of this ocean is clearly defined is uncertain. It may simply blend into the nearly liquid atmosphere.

As *Voyagers 1* and *2* flew close to Jupiter in 1979, they discovered two interesting luminous phenomena in Jupiter's

---

Shortly after sunset, hovering in the cloud tops far into Jupiter's polar regions, an eerie auroral display flutters its multicolored curtains (left). *Electrically charged particles, having made the three-quarters-of-a-billion-kilometer journey from the sun, are sucked into Jupiter's powerful magnetic whirlpool. They then slam into molecules of the atmosphere, which begin to glow like the gases in neon lights. The aurora hisses, crackles, and flickers like a garish sign in front of a seedy hotel.*

The orange sphere is Io, lit by the setting sun, its dark face dimly illuminated by Jupiter light. Io appears three times the size of our moon in Earth's sky.

atmosphere. First, ghostly auroras play high over Jupiter's polar regions. Like Earth, Jupiter has Van Allen belts of ions, or charged particles from the sun, trapped in its magnetic field. These can't easily escape, but can move along the magnetic field toward either the north or south magnetic pole of Jupiter. As a result, swarms of ions sometimes crash into the upper atmosphere, colliding with air atoms and molecules and making them glow with strange colors. These colors depend upon the types of atoms involved and energies of their collision. The auroras of both Jupiter and Earth are formed in the same manner.

The second discovery was of a pattern of bright glimmers on time exposure photos of Jupiter's night side. These are believed to be lightning blasts in the clouds. As on Earth, turbulent cloud motions probably cause atmospheric charges, leading to eventual discharges in stupendous lightning bolts.

Decades before the 1979 *Voyager* flights, a few astronomers had raised the possibility that Jupiter might have a slight ring, too thin and faint to show up from Earth. Generally, this idea was ignored because it couldn't be proven at

that time and wasn't required by any proposed theories of planet formation. Nonetheless, as *Voyager* approached Jupiter, NASA scientists planned to take some pictures of the region near Jupiter to check for possible rings or faint inner moons. It was a long shot, but it paid off beyond their wildest dreams. The first pictures, and several subsequent views, revealed a beautiful narrow ring, brightest at its outer edge. It is clearest when viewed from the far side of Jupiter, backlighted, with sunlight coming through it toward the cameras. Primarily composed of a swarm of microscopic particles, it looks rather like a thin cloud of cigarette smoke—dim in the middle of a room with the light behind you, but bright where it hangs in front of a lamp with light shining through it.

The same photos also showed a small satellite, previously unknown, at the outer edge of the newly discovered ring. The implication is that the ring is fed by tiny particles that get blasted off this moon by small meteorites. Dynamical studies show that these particles spiral in toward Jupiter, thus augmenting the ring. The ring seems, then, to be not a static feature, like a rock or a hill, but rather

dynamic, like a river. The ring we will see years from now will be different from the one we see today, because new material is constantly flowing through it.

Jupiter remains one of the most alien wildernesses in the solar system—a reminder of the awesome environments that nature can create.

Jupiter's satellite system to scale. Left to right: Sinope, Pasiphae, Carne, Ananke, Lysithea, Elara, Himalia, Callisto, Ganymede, Europa, Io, Amalthea.

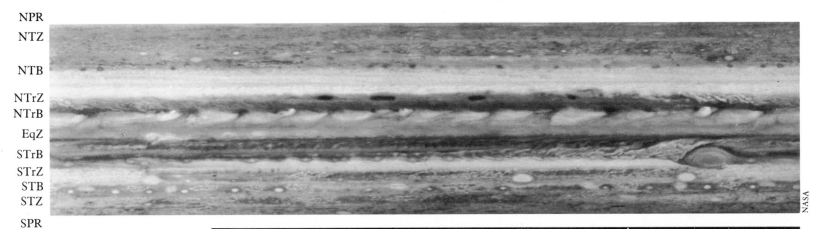

NPR

NTZ

NTB

NTrZ
NTrB
EqZ
STrB
STrZ
STB
STZ

SPR

Jupiter, which has no permanent features, is divided into horizontal zones and belts. From top *to* bottom: *North Polar Region, North Temperate Zone, North Temperate Belt, North Tropical Zone, North Tropical Belt, Equatorial Zone, South Tropical Belt, South Tropical Zone, South Temperate Belt, South Temperate Zone, South Polar Region.*

Little Sinope (right) *hangs in space 24 million kilometers (14,880,000 miles) from Jupiter. The giant planet and its four major moons, the Galilean satellites, can be easily seen, looking very much as they do in a backyard telescope back on Earth. Sinope's orbit, however, is inclined to Jupiter's equator, swinging high above and below the planet. Here we're looking "down" on the Jupiter system from a higher angle than we can get from Earth. Sinope is probably a ball of rock and ice, barely 20 kilometers (12 miles) across. Its small size, eccentric orbit, and dark surface may indicate that it was originally a wandering asteroid captured by Jupiter's gravity*

SATURN LORD OF THE RINGS

Average distance from the sun: 1,427,000,000 km.
Length of year: 29.46 years
Length of day: 10 hours 40 minutes
Diameter: 120,000 km.
Surface gravity (Earth=1): .93
Composition: hydrogen; ices; hydrogen atmosphere

Saturn is a smaller-scale, colder, and calmer version of Jupiter, most famous for its rings. We are so used to the verbal and visual picture of "Saturn, the ringed planet" that it is hard to focus on the planet itself—the yellowish-tan ball that floats in the center of the ring system.

A brief examination of the size and mass of Saturn reveals that it is the least dense planet in the solar system. In fact, it is less dense than water; if there were a large enough ocean, Saturn could float in it. Saturn, like the other three giant planets, is rich in the lightest-weight element, hydrogen. Like Jupiter, it is believed to have a small, rocky core at its center, several times the size of Earth, surrounded by liquid metallic hydrogen. The outermost layer may be an ocean of liquid hydrogen and other compounds.

Saturn's clouds, viewed from space, lack the vivid colors and contrasting swirls of Jupiter's, but they do have the same general arrangement of dark bands and lighter zones paralleling the equator. Close-up photos by the *Voyager* spacecraft in 1980 showed some turbulent swirls, and telescopic observers from Earth see occasional dark and light disturbances that erupt for some months before fading. As on Jupiter, the main two dark bands straddle a lighter equatorial region.

Saturn, like Jupiter, radiates measurably more heat from its interior than it receives from the sun. This heat flow may help stir the cloud patterns in its atmosphere. The reason for the lower contrast and seemingly lower activity of Saturn's atmosphere may involve a high level of haze overlying the main cloud decks.

Saturn has the grandest ring system of all the planets. Unlike Jupiter's, Saturn's rings are very prominent even as seen from Earth through small telescopes.

The particles of Saturn's rings are larger than those in Jupiter's rings.

Spectroscopic observations in the 1960s proved that the ring particles are composed of (or at

Saturn's rings (below) *swing around the crescent planet (overexposed here to bring out ring detail). The transparency of the C and A rings is easy to see, as well as the opaque areas of the B ring.*

NASA

23

SATURN

least coated with) frozen water. And other testing methods, such as the bouncing of radar signals off the rings, indicate that the dominant particles range from the size of marbles to basketballs; larger ones, though more rare, may reach sizes as large as 1 to 100 kilometers (.6 to 60 miles) in diameter. These larger particles (or moonlets) orbit among the swarm of smaller particles and may possibly determine the distances between them. Observations from Earth and space vehicles in 1969–80 revealed several small moons, on the order of forty to two hundred kilometers across, orbiting at the very edge of the ring. Gravitational forces from these moons and also from the more distant satellites of Saturn are important in further control of the edges and divisions of the rings.

Although early observers had little physical information about Saturn and its rings, they were able to perform many interesting geometric studies of the ring configuration. As early as 1865, the British observer Richard Procter realized that the sky of Saturn, at least from the level of the cloud tops, must present an extraordinary sight, with the ring system arching through the

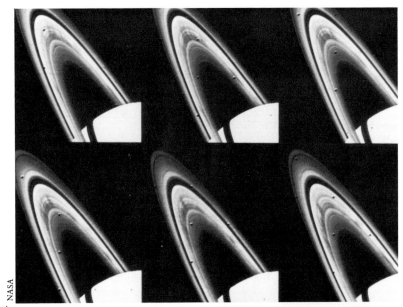

NASA

Mysterious dark spokes appear in these Voyager 1 *photographs* (above). *They were taken in sequence, from upper left to lower right.*

firmament like a midnight rainbow.

The rings are tipped relative to Earth's position, and as Saturn orbits around the sun we see them from different angles. This varying perspective reveals that they are extremely thin and flat; they are about 275,000 kilometers (170,500 miles) in diameter, but their thickness is probably less than one kilometer and may be only a few hundred meters. Visiting the rings is quite a feat. If we fly into the Saturn ring system at normal spacecraft speeds (i.e., on an unpowered

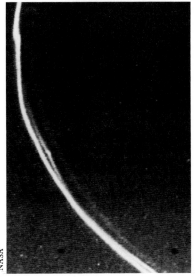

NASA

The enigmatic F ring and its inexplicable lumps and braids (above).

A RING
Cassini's
Division
B RING
C RING

F RING

orbit falling toward or around Saturn) we will pass through the rings in less than a tenth of a second unless we approach at a very shallow angle to the ring plane. And if we do that, we will be within the rings for a long time, and stand a fairly good chance of hitting a dangerously large body.

But if we slow down, using retro-rockets, and move into the rings, what a sight we'll see! In the most heavily populated parts of the rings, much of the sunlight is blocked out by whirling bodies that cause a thousand twinkling eclipses. Clouds of hailstones spin by, snow-white and glittering. And the sky is filled with passing bodies stretching off to a strange horizon—the distant vanishing point defined by the ring plane . . . a virtual highway in the sky.

Earthbound telescopic observers have long known that the rings were structured. In the 1600s French astronomer Jean

A view never seen from Earth: a crescent Saturn (above), *photographed by* Voyager 1.

NASA

Dominique Cassini discovered a prominent gap. Other finer gaps, called *divisions*, have also been seen. Cassini's Division forms the boundary between the outer ring, called the *A Ring*, and the inner *B Ring*, which is somewhat brighter than its counterpart. Inside the B Ring is a much fainter division of rings called the

---

Saturn's satellite system to scale. Left to right: Phoebe, Iapetus, Hyperion, Titan, Rhea, Dione, Tethys, Enceladus, Mimas. There are several small moons between Mimas and Saturn's rings.

C Ring. Other ring portions have been tentatively identified by various observers, including the so-called D, E, and F Rings.

But when *Voyager* flew by in 1980, it revealed a vastly more intricate structure than was explained by the old theories. For example, in one of the obscure divisions inside the rings, a very narrow ring was found that was not circular, but rather elliptical in profile. On one side of Saturn it seemed to be centered in its division, but on the other side it was well off-center, toward the edge of the division. This finding may help explain some very puzzling observations: various telescopic observers have drawn pictures of divisions in different positions in

the rings at different times. Perhaps changing configurations of satellites actually change positions of ring divisions. Or perhaps rotation of the rings brings different structures into view.

This last possibility seemed to be established by *Voyager* when it discovered strange, radial streaks in the rings that could be seen rotating around Saturn in time-lapse movies made by *Voyager* cameras. The explanation of these spokes, similar to markings sketched by observers as long ago as the 1880s, is also unknown. Since the particles in any ring system move at different speeds (those closer to a planet move more rapidly than those farther from it), it seems impossible for a radial pattern to be maintained, since the part of the spoke closer to Saturn should be moving more rapidly than its outer end.

The complexity of the rings was further made apparent by *Voyager*'s discovery of an intricate structure of filaments, seemingly braided in appearance, in a narrow ring just outside the A Ring. How swarms of ring particles can arrange themselves in filaments and tangle with neighboring filaments is unknown.

Many scientists believe that Saturn's rings originated from

Saturn (overleaf) *is a world of genuine magic, circumscribed by the compass of a mythic draftsman. Hovering above dark, wind-torn clouds at a latitude that, on Earth, would put us over Mexico City, the great rings cover the southern sky like the rainbow bridge Viking heroes took to Valhalla. In this nearly 180-degree view, they sweep from the east, at our extreme left, to the west, where the shadow of Saturn itself has bitten off a reddened mouthful.*

The sun will be rising soon. For all of its enormous size (nine and a half times the diameter of Earth), Saturn rotates in only slightly more than ten hours—so rapidly that the planet bulges strikingly at its equator and objects there weigh one-sixth less than they do at the poles. The night has lasted scarcely five hours, Saturn's shadow sweeping across the rings like the hand of a clock.

Although softened by the atmosphere, much of the fine detail that makes the rings look like a phonograph record still comes through. Many of the largest features—Cassini's Division, the dark band at the top, or the French Division between the bright B Ring and the inner C Ring—are fairly stable and permanent. Observations over the past 150 years or so would seem to imply changes in many of the finer details: they come and go, passing through the rings like ripples in a circular pond.

Ron Miller

Saturn's Ring Panorama

SATURN

material left over in the vicinity of the planet after Saturn formed. The rings of Saturn, like the rings of Jupiter and Uranus, are mostly located within a critical distance from the planet called *Roche's limit*. This is the distance within which the stretching forces associated with a planet would break a large, weak object apart. Just as the moon raises tidal forces in the oceans of Earth, any planet can stretch and raise bulges on the surface of a satellite.

Most satellites are outside Roche's limit, and therefore are too far away for tidal forces to do any structural damage. However, inside Roche's limit, this stretching can pull a weak moon to pieces. Therefore, if small particles are formed inside this zone, they are unable to aggregate into a single satellite.

Some scientists believe that the Saturn ring material may have been left from primeval days because of this effect. Others suspect that a large comet or asteroid may have passed so close to Saturn that it broke up inside Roche's limit, spewing a cloud of ring particles into the area. Alternatively, a small moon located within or near Roche's limit may have been shattered into many pieces by the

Ron Miller

Seen from Hyperion, Saturn (left) is nearly three times farther from us than Earth is from the moon. Still, the ringed planet appears ten times larger than our moon does. Only 257,600 kilometers (159,712 miles) away is the giant moon Titan, looking like an enormous tangerine. Saturn, here, is tilted to almost its greatest extent in relation to the sun. The shadow of the rings covers nearly the entire southern hemisphere. Light reflected from the brilliantly white rings dimly illuminates the planet's night side.

Hyperion is one of Saturn's smallest satellites, only 290 kilometers (180 miles) in diameter. It is the third from the outermost moon, between Titan and Iapetus. Its surface is bright and probably contains some frozen water and other ices in thickets of large crystals, or perhaps a sheen of frost or blanket of snow.

To the west, just before sunset, Saturn's rings are a pale rainbow arching high into the sky (right). A halo surrounding the offstage sun makes a smaller loop. The rings provide an endlessly changing spectacle as the planet's shadow sweeps over them.

Ron Miller

29

SATURN

impact of a large meteorite.

Whatever the origin and history of Saturn and its rings, the system reminds us that nature cannot only create bizarre and beautiful environments on planetary surfaces, it may create astonishing environments in orbit around those worlds.

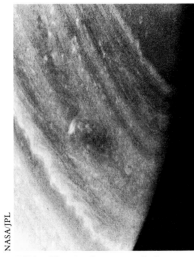

NASA/JPL

Ribbonlike clouds are stretched across Saturn by 800-kilometer winds (above)

Ron Miller

The sky is split in two by a bright crack (left). We are at Saturn's equator, directly below the rings, and are seeing them edge on. In spite of their enormous width, the rings are very thin— probably less than a kilometer wide. From here, 14,000 kilometers (8,680 miles) away, they appear to be the width of a pencil line.

The narrowness of the ring's shadow, seen to the left, indicates that it is almost spring leading to a fifteen-year winter in one hemisphere, and a fifteen-year summer in the other.

URANUS A SIDEWAYS WORLD

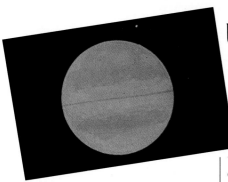

Average distance from the sun:
2,870,000,000 km.
Length of year: 84 years
Length of day: 15.5 hours (?)
Diameter: 51,800 km. (?)
Surface gravity (Earth = 1): .8
Composition: ice, hydrogen, helium;
hydrogen-methane atmosphere

Uranus is the first full-fledged planet to have been discovered since prehistoric times. The discovery was made by the English astronomer William Herschel in 1781, while he was charting stars with his six- inch telescope.

Uranus is so far from Earth that it presents only a tiny disk in our largest telescopes. No markings are prominent, but the disk shows a pale greenish or greenish-blue tinge. This color may arise when yellowish or tan clouds, like those of Saturn, are seen through a deep haze layer that scatters bluish light like the light of our own sky. Photos in specially selected wavelengths reveal that the cloud belts and zones run parallel to the equator, as on Jupiter and Saturn.

Basically, Uranus seems to be a scaled-down version of those two planets, but with a few similarities to Earth. For example, its diameter is only four times that of Earth; Jupiter's is eleven times that of Earth. Its density lies between that of Jupiter and Saturn (which are

less dense) and Earth and the moon (which are more dense). Some theoretical studies suggest that most of Uranus's interior is a mixture of about half water, one quarter methane and one quarter rocky material. Theoretical studies also show that Uranus probably has a core of rocky and metallic material that resembles Earth, but is slightly larger. This core is believed to be overlaid by a layer of ice—frozen methane, water, and ammonia—which in turn is covered by a vast ocean of liquid metallic hydrogen and perhaps a surface ocean of ordinary liquid hydrogen. Above this is the deep atmosphere.

We may learn more about Uranus in 1986, when the *Voyager 2* spacecraft flies close to the planet.

On March 10, 1977, a number of astronomers convened in Africa and near the Indian Ocean to watch Uranus pass in front of a relatively bright star. The plan was to monitor the starlight as the star disappeared behind Uranus's atmosphere and reappeared on the other side.

Such observations were meant to provide additional information about the atmospheric composition and structure of Uranus.

The astronomers were in for a surprise. The star went through a pattern of brief dimming *before* Uranus passed in front of it and then repeated the same pattern in reverse on the other side. Something that existed on both sides of the disk, in a uniform pattern, was blocking the starlight. This "something" was surmised to be a series of narrow rings around the planet. Later observations with other types of instruments have confirmed the existence of Uranus's rings.

These rings are very different from Jupiter's or Saturn's. Jupiter's ring looks like a single, moderate-width band. Saturn's rings are a wide system divided by more than a hundred narrow gaps. Uranus, in contrast, has several narrow rings separated by

Uranus's satellite system to scale. Left to right: *Oberon, Titania, Umbriel, Ariel, Miranda.*

32

wide gaps. Their structure is believed to relate to forces exerted on the rings by nearby satellites, possibly located within the ring system. The exact physical mechanism that causes the spacing of the rings is still uncertain. Some theorists suggest that each Uranian ring is associated with a small, as yet undiscovered moonlet located in or near the ring.

A distinctive feature of Uranus and its ring system is that the whole system is "tipped over." Most planets' equators lie nearly

Unlike the brilliant rings that encircle Saturn, Uranus's are narrow, dark hoops (left). They are barely visible until you are virtually within them—as we are here. We are positioned near a small moon that orbits between two of the nine rings. One of these moons may be within each gap, responsible for keeping these spaces swept clear of dark rock and dust. The bright disks are Ariel and Miranda, the innermost two of Uranus's five moons (not counting the moonlets within the rings). The shadows of two outer rings stripe our little moon, and the shadows of all nine rings on the planet itself run just below the fuzzy line of rubble.

While Uranus's rings lack the almost mystic grandeur of Saturn's, we can appreciate their simplicity: they are drawn like a vast geometric exercise around the placid, turquoise planet.

in the plane of the solar system, with their polar axes of rotation pointing "up" or "down," nearly perpendicular to the plane of the solar system. Jupiter's equator and rings, for instance, are only a degree or so off the plane of the solar system. Earth is 23.5 degrees off, meaning Earth's northern hemisphere gets more direct sunlight for part of the year (summer), and less sunlight six months later (winter). The seasons thus arise because Earth's northern hemisphere is tipped toward the sun for half the year and away for the other half.

In contrast, Uranus's pole lies almost in the plane of the solar system. This means that Uranus has extraordinary seasons. For one quarter of the Uranian year, the north pole gets direct sun.

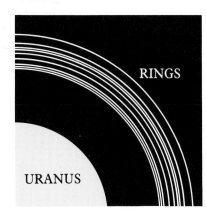

RINGS

URANUS

During this season, the equator is in perpetual twilight. After a quarter of its trip around the sun is completed, we find the sun rising and setting over the equator during each of Uranus's eighteen-hour days; the north pole at this time is just grazed by light. After another quarter of a year, the northern hemisphere is deep in night. The north pole's night lasts half a "year." After another quarter year, the sun is over the equator again and summer begins to return to the north hemisphere.

On Earth, the north (or south) pole is plunged into night for half the year, but on Uranus, this long winter night happens to much of each hemisphere. At the height of summer, the sun shines continually almost directly above the north pole. Because Uranus takes 165 years to go around the sun, the night at the pole lasts a frigid, black 82 years, followed by a polar "day" equally as long.

Other planets are illuminated by sunlight that comes only from "the side", i.e., from a direction nearly in the equatorial plane. Uranus can be lit from almost any direction, above the equator or above the pole. Uranus's rings lie around its equator, so they too can be lit edge on, and sometimes from directly above or below.

The changing phases of the globe, the ever-shifting light on both Uranus and the rings, and the slipping of the ring's shadow back and forth across the globe present an endlessly changing panorama to observers in nearby space or on one of Uranus's moons.

Above Uranus, its cloud-tops fogbound in a methane haze (overleaf), we can get a clear view of its spidery system of rings. We're looking "down" on them from latitude 45 degrees N—if Earth had rings identical to these (but scaled down to match its smaller size), this is what we'd be seeing as we hovered over Ottawa or Portland. In order to see their full width, we need this wide-angle picture—spanning about 135 degrees, more than you could see with the naked eye without turning your head.

The rings are very dark, even though the sun is shining on them. Unlike the rings of Saturn, which are made of billions of large and small particles all coated with (or made of) ice, Uranus's rings are the color of powdered coal. Tiny moonlets, which may orbit within the gaps between the rings, may keep these areas swept clean of debris. The sun is nineteen times smaller than in Earth's sky, providing less than one three-hundred-and-fiftieth the amount of light and heat.

Ron Miller

Above Uranus and Rings

Ron Miller

From an orbit close to icebound moon Umbriel, Uranus (above) reveals one of its unusual-looking phases caused by its unique axial tilt. Inner moons Ariel (the larger one) and Miranda share the same phase. Uranus is on its way from its northern hemisphere midsummer (when its north pole is pointed directly at the sun) to spring, when the terminator (the line separating night from day) is at a right angle to the equator, like a "normal" planet. Uranus's rings are in the plane of its equator, like Saturn's and Jupiter's. Like those planets, the moons of Uranus are probably composed mainly of ice.

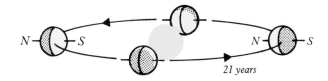

Uranus's axis is tipped nearly 90 degrees; twice during its year, one of its poles points directly at the sun.

NEPTUNE THE LAST GIANT

Average distance from the sun:
4,497,000,000 km.
Length of year: 165 years
Length of day: 18 hours 30 minutes (?)
Diameter: 49,500 km. (?)
Surface gravity (Earth=1): 1.15 (?)
Composition: ice, hydrogen, helium;
hydrogen-methane atmosphere

Neptune was discovered telescopically in 1846 after calculations showed that Uranus's motions were being disturbed by a still more distant planet. Neptune is too far away to be seen by the naked eye, but when dynamicists calculated the orbit that seemed to be required by the hypothetical planet's effects on Uranus, it was soon located telescopically.

The night of space gets darker as we move out from the sun. In the outermost region of the solar system, planets are far apart and sunlight is dim. Our knowledge of this area is slight, and Neptune remains something of an enigma. It is so far from Earth that it appears only as a tiny, nearly featureless disk even in the most powerful telescopes. To date (1980), no rings have been detected, but the recent discovery of faint rings around Jupiter and Uranus, together with the great ring system of Saturn, make us wonder if Neptune, like its three giant brothers, doesn't possess rings as well.

Neptune's atmosphere is rich in methane and hydrogen, and probably contains other gases as yet undiscovered. A cloud layer similar to those of Jupiter, Saturn, and Uranus is believed to surround the planet, probably with belts and zones arranged parallel to the equator. However, a bluish tinge reported by many visual observers suggests that the clouds may be overlaid by a thick, hazy layer of gas that scatters blue light like our own sky does.

Neptune's interior structure is probably similar to that of Uranus. Theoretical studies suggest a rocky core several times the size of Earth's core surrounded by massive layers of ices. These are probably covered by a deep ocean of liquid molecular hydrogen. If the instruments of the *Voyager* spacecraft last long enough, it may be possible for one of the *Voyagers* to fly on past its planned rendezvous with Uranus to reach Neptune; and the day may soon come when we have a clearer understanding of this remote giant.

Deep in the frigid, hazy atmosphere of Neptune (right), we watch its giant moon Triton about to eclipse a tiny, chilly sun. Yellowish clouds—similar to those that blanket Jupiter and Saturn—roil sluggishly. Neptune is so cold that what little weather there is stirs slowly in an icy torpor. Most of the atmosphere surrounding us is hydrogen and noisome methane—"swamp gas." A thick ocean of haze surrounds us. It is fairly clear and scatters blue light like Earth's atmosphere does. The sight of the yellowish clouds through this blue haze explains the slightly greenish tint Neptune has when seen from space or through Earthbound telescopes. Triton appears almost half again larger than Earth's moon, while the shrunken sun is reduced thirty times in diameter: eight hundred times less in area, luminosity—and warmth.

Neptune's satellite system to scale. Left to right: Nereid, Triton.

Ron Miller

EARTH THE PLANET OF LIFE

Average distance from the sun:
  149,600,000 km.
Length of year: 365 days
Length of day: 23 hours 56 minutes
Diameter: 12,756 km.
Surface gravity (Earth = 1): 1
Composition: nickel-iron, silicates;
  nitrogen-oxygen atmosphere

At the turn of the century, rocket pioneer Konstantin Tsiolkovsky said, "Earth is the cradle of mankind, but one cannot live in the cradle forever." Earth is more than a cradle to us; Earth has helped shape us. And we are only its most recent product. Geologic studies have shown that Earth has evolved dramatically since it formed 4.5 billion years ago. It has gone through changes that have been extremely rapid, and others that have been very slow.

Earth's atmosphere had much less oxygen for the first billion years of its existence, as shown by oxygen-poor sediments deposited during that period. The amount of oxygen increased after plant life evolved. This apparently happened early on; lab experiments confirm that complex organic molecules form easily as a result of lightning strokes and other bursts of energy in the primordial atmosphere.

For the first half of Earth's history, little life could be found on land, and there were few organisms even in the sea. Although signs of ancient life as old as 3 billion years have been found in rocks, trilobites and other organisms large enough to leave obvious fossils did not evolve until 600 million years ago —i.e., after 87 percent of Earth's history had already elapsed. The modern continents didn't start to take shape until 300 million years ago, and dramatic environmental changes have occurred even in the last 100 million years.

The point of all this is that Earth has been constantly changing. It may always have been a blue planet with swirling white clouds, but it has not always been a passive stage on which we living creatures play out our roles. The stage itself has changed, and so have we. Fortunately, we evolved fast enough to keep up with our changing Earth. But today, we have the technological ability to alter Earth's environment faster than we can evolve, and we threaten ourselves and other species in the process.

As environmentalist René Dubos has pointed out, our connection to Earth is umbilical. Earth has nourished us, and we are suited to it and comfortable

NASA

North America from space (above).

with it. Writers such as Carl Sagan stress the beauty of Earth; but this beauty is not an intrinsic or universal absolute. Earth is beautiful to us because we have evolved on it and have adapted to it; it is the only place in the solar

The good spaceship Earth is 25,000 kilometers (15,500 miles) ahead of us (right). Cloud-marbled, looking like a blue-and-white bowling ball, is it only bias that makes it seem so beautiful and appealing to us? It is swarming with life brought about by many happy circumstances. For example, it is just the right distance from the sun. If it were a bit farther out, the swirling patterns of water vapor would be frozen on the surface; even farther, the continents would be buried beneath crawling glaciers; farther still, the precious atmosphere itself would lie in drifts and frozen lakes. If it were closer to the sun, it would be an arid inferno like Venus. In the scale of this picture, the atmosphere is as thin as the ink with which it is printed. And as far as anyone knows, all the life there is in the universe exists within this fragile film.

The aurora borealis flickers in a ragged ring around the north pole while the distant moon wanly lights the night side of our home planet. So large is our moon in proportion to Earth (only Pluto and its moon Charon are as close in size) that it would be reasonable to call Earth and the moon a double planet.

39

The rugged, snowcapped mountains of Greenland, near the shore of the ice-filled Greenland sea (below).

NASA

Ron Miller

As Earth grew older, its crust buckled and wrinkled. Continental plates collided like sheets of ice in a frozen river. Where plates met, where the crust split and faulted, mountains were born (left). Volcanoes laboriously built up miles-high cones or broad piles of ash and cinder, layer by layer.

The collision of India and Asia created the Himalayan plateau; one of the wrinkles on top of it became Mt. Everest. An enormous block of Earth's crust was suddenly vertically thrust 3,000 meters (9,900 feet). The craggy, eroded remnant of this vast cliff are the Tetons.

Earth's largest mountain was created by a volcano. From its underwater base to its summit, Hawaii's Mauna Kea is 10 kilometers (6.2 miles) high, 1,500 meters taller than Everest.

Precipitous ranges like the Rocky Mountains, the Alps, or the Himalayas are relatively young mountains. But mountains, too, grow old. Millions of years of wind, rain, and plant life can eventually reduce a range like the Andes to the elderly undulations of the Appalachians.

system where we can stand naked, breathing the air and feeling the sun, and see water trickle by our feet.

Space exploration is helping us to realize that Earth is a fecund paradise. That realization should provide us with enough motivation to try to keep it that way. How beautiful and how varied Earth is! Its active geology has created landscapes of enormous variety—rolling blue seas of liquid water, windy ravines, fields of sand dunes, dark overgrown jungles, barren polar deserts, volcanic explosions, golden prairies, waterfalls, and jagged peaks looming over glacial valleys. Our skys encompass clear blue vistas, white cloud puffs, dusty yellow glows, sunset reds, foggy grays, crystal moonlight, and spectral auroras.

Today there is little doubt that nomadic humans will eventually be able to journey to other planets, build space cities, collect energy, manufacture new tools, and multiply. There is increasing evidence that this may be a way to save our race from some natural or man-made disaster that could make Earth uninhabitable. By moving some of our industries to space, we may be able to reverse the rising tide of acid rains, nuclear waste production, desperate strip mining for fuels and oils, hydrocarbon damage to the ozone layer that protects us from ultraviolet rays, and other

results of the activities of a population whose appetites have grown beyond Earth's ability to sustain them.

Space exploration will also help us to understand whether Earth is so unusual that life is rare in the universe, or whether Earth-like conditions, and hence living organisms, are common.

Nineteenth-century beliefs that nearby planets were populated are both overly anthropocentric and incorrect. Recent space flights suggest that we don't share the solar system with anyone else. Certainly, there are no civilizations on Mars, no dusty cities or even organic sediments. Nonetheless, an

After leaving the sun, particles blown from a sunspot slam into Earth's upper atmosphere, creating an aurora (right) *in exactly the same way that electricity passing through the gas in a neon tube causes it to glow. Different gases give off different colors. Earth's magnetic field funnels the charged particles into the north and south magnetic poles like water funneled into a drain. Here, not far from the magnetic north pole (which is not the same as the north pole of Earth's axis—that is nearly 1,600 kilometers or 992 miles, farther north), the shimmering atmosphere is alive with ghostly curtains and veils. They constantly and abruptly flicker and change, as though they were connected to a switch. In the crystalline silence of the subfreezing air, we can even hear the aurora, faintly hissing and crackling like an old radio.*

Ron Miller

41

EARTH

Ron Miller

The great sand dunes of Africa's Namib Desert dominate this photograph (below), made from an altitude of 914 kilometers.

NASA

Fully a third of Earth's surface is desert (left)—more, if the Arctic and Antarctic regions are included. There are almost as many definitions of desert as there are deserts. All are arid, with little or no rainfall; there is a desert on the west coast of South America that has never known any rain. Even the polar regions are arid, receiving scarcely ten inches of precipitation a year. Still, lifelessness is not necessarily a characteristic of deserts. The most hostile of Earth's arid regions, seemingly without a trace of water, encrusted with salts and broiling in temperatures reaching 48 or even 65° C (120 or 150° F) can still support life, and, in some cases, that life even flourishes. The danger of the deserts is that they are expanding, as we destroy more and more of our land and its delicate ecological balance. Three-quarters of this planet is underwater, a third of the remainder is desert, and a large portion of what's left is uninhabitable or not useful for one reason or another—mountains, swamplands, etc. We have very little planet to waste.

occasional meteorite may contain the building blocks of life—amino acids—formed in that meteorite's original environment, which was probably beneath the surface of an asteroidlike parent body somewhere in the solar system.

A comparison of data from both Earth and the planets originally suggested that the evolution of intelligent life required a relatively stable environment in which organisms could evolve over a long period. But terrestrial and planetary research has shed new light on this idea in recent years. Geologists now recognize from fossil records that there were some abrupt changes in the types of plants and animals on Earth. Recent work suggests that these abrupt shifts (for example, from dinosaur to mammal) were not due to changes in the rate of biological evolution or slow changes in climate. Instead, they seem to reflect sudden changes in the biological environment.

To take a simplified example, during the period when dinosaurs dominated most regions of Earth, proto-mammals adapted to certain "backwater" regions—perhaps mountain valleys—where the dinosaurs could not compete. Suddenly, the climate changed; some evidence points to a sudden onset of colder winters. Vegetarian dinosaurs had neither the food nor the constitution needed to get through these

winters, and when they died out, so did carnivorous predators. On the other hand, species of warm-blooded mammals were ideally suited to the new climate and spread out from their earlier enclaves.

What changed Earth so suddenly as to spur evolution in this direction? Recent studies of soil sediments have produced an astonishing answer. In 65-million-year-old strata, exactly at the layer corresponding to the change in fossils from dinosaur to mammal, scientists have found increases in the amounts of iridium and other rare elements. The enrichment pattern matches elements known to be abundant in certain meteorites. In 1980, several groups of geoscientists therefore announced a new theory, that the sudden climate change a million years ago was caused by the impact of a giant meteorite—an asteroid estimated to be ten kilometers (six miles) across. It may have thrown up enough dust to screen out sunlight, cooling the atmosphere enough to lead to the decline of dinosaurs.

Earth is one globe out of many, with its own environmental evolution. Our soil is planetary material, evolved (like us) from dust that condensed in a cloud near the sun 4.6 billion years ago. Earth is the environment whose changes, one step at a time, led to the appearance of human beings. It is obvious,

EARTH

The varying depths of the ocean are apparent in this Skylab 3 *photograph of the Great Bahama Bank (below).*

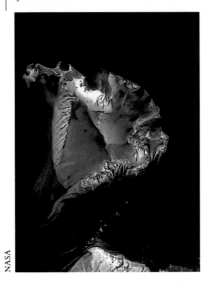

NASA

This is a typical Earth "landscape" (overleaf), the one an extraterrestrial visitor has a 75 percent change of seeing before any other. It belies the name we have given our planet and illustrates how careful we must be in extrapolating the nature of a planet from a single, random sample. Certainly, what we consider the most important area of Earth is the remaining 25 percent—and only a small portion of that. But the seas of Earth dominate more than a mere area: they dominate our weather and climate, and ultimately the fate of the continents themselves. They nurtured life itself, and maintain it. Perhaps we should have named our planet Ocean.

Earth's Seas

EARTH

Ron Miller

EARTH

The coexistence of liquid water and a solid landscape is Earth's most striking feature (left). Here, at the boundaries between Earth's oceans and its smaller patches of land, we see clues to our genesis.

During most of Earth's history, life was concentrated in the seas, and it vigorously emerged onto land only in the last billion years. This coastal scene shows some seeds of that process. Rocks in the lower right are covered with greenish patches of lichens or moss, which may be similar to some of the simple early land plants. Along the shore to the left are some clumpy organisms called stromatolites. *These cabbagelike plants are among the earliest known; they developed on seacoasts over a billion years ago. In the millennia since the first stromatolites appeared, the increased oxygen content of the air and other changes have made today's Earth less suited to them than it was two or three billion years ago. Hence, stromatolites are rare today, found principally in saltwater marshes in coastal areas of Australia and Baja California. But they, and the beautiful seacoasts of Earth, bear witness to the emergence of the solar system's only life forms, crawling from sea waters onto dry land.*

An erupting volcano in the Galápagos Islands (below) leaves a long plume, seen from over 430 kilometers up.

NASA

therefore, that humans are uniquely adapted to Earth, and that Earth should be precious to us. At the same time, Earth is one small planet, subject to cosmic forces. Its climate is subtly influenced by solar radiation, which flickers both on a short time scale of days and a long time scale of many years. Evidence suggests that droughts and other climate variations are correlated with these flickers. Earth is battered by meteorites and occasional asteroids, but it is also so geologically active that erosion removes the crater scars that accumulate on smaller, less active worlds.

With cosmic perspective, we can better appreciate the many environments on Earth. There is beauty not only in warm pools and leafy glades, but also in the dark currents of the sea floors, the ripples of desert dunes, and the silent play of fluorescent auroras over icy polar wastelands. As our eyes shift from these extreme regions of Earth to the other planets, we can recognize kindred landscapes there. Earth is our birthplace and our home, but, to paraphrase Tsiolkovsky . . . we can't expect to live in the same house forever.

Don Davis

Stripped of its oceans, Earth becomes an alien world, looking much more like one of the other planets. Water makes our world unique.

The Earth and its moon compared in size.

The Earth-moon distance to scale.

Fooled into thinking night has fallen, an owl is silhouetted against the pearly streamers of the sun's corona during a total eclipse (left). Seen from Earth, our moon is almost exactly the same size as the sun (it sometimes looks a little smaller, depending upon where it is in its orbit). By this happy coincidence, only the solar disk is covered when the moon passes before it: none of the fine details and structure of the sun's atmosphere are hidden.

At the predicted hour, we see the moon carve a dark segment cut out of the edge of the sun, slowly advancing from the west, eating away at the sun until only a thin, brilliant crescent remains. At the moment the sun disappears, a wan and sinister twilight replaces the bright daylit sky. We can even see stars, and Venus shines above the eclipsed sun in the noon sky. The landscape around us remains vaguely illuminated—the solar corona provides as much light as a full moon. A total eclipse is a rare, sublime, and brief spectacle; at its longest, it lasts less than eight minutes.

The sun glints off the ocean near East Africa in this Apollo 11 view of a crescent Earth.

VENUS THE VEILED INFERNO

Average distance from the sun:
108,200,000 km.
Length of year: 224.7 days
Length of day: 5,832 hours (243 days)
Diameter: 12,100 km.
Surface gravity (Earth = 1): .91
Composition: nickel-iron, silicates;
carbon dioxide atmosphere

Closest to Earth in distance and size, Venus at the same time is one of the most unusual and fearsome environments in the entire solar system, with a dense carbon dioxide atmosphere and a surface temperature of 460°C (865°F). Because Venus formed not too far from Earth and has a similar mass, it probably has a similar basic mineralogy. Why, then, the differences?

For a start, Venus's proximity to the sun means that it probably contained less water to begin with; certainly it never had the great oceans of Earth. Long ago, when the interiors of Venus and Earth heated up due to radioactivity, volcanoes and fumaroles erupted. Carbon dioxide was one of the most abundant volcanic gases. On Earth this carbon dioxide dissolved rapidly in the oceans, making weak carbonic acid that reacted with the rocky ocean floor, ultimately creating carbonic rocks. Most of Earth's vast amount of carbon dioxide thus "disappeared" into the oceans and rocks, and did not accumulate in the atmosphere.

But on Venus there was no ocean. Venus contains the same amount of carbon dioxide as Earth, but on Venus, all of this very heavy gas is in the atmosphere. Venus, therefore, has a very dense atmosphere, exerting more than ninety times as much pressure at the surface as Earth's does. Instead of a familiar 14 pounds per square inch, Venus's air pressure is a crushing 1,260 pounds per square inch! Standing on the surface of Venus, you are subject to a pressure comparable to being over 1,000 meters—over half a mile—beneath the surface of the sea.

Venus's excessive carbon dioxide had another consequence, called the *greenhouse effect*. The glass windows of a greenhouse let in sunlight, but prevent infrared "heat waves" from getting back out. The infrared radiation is thus trapped, warming the air inside. Similarly, carbon dioxide gas blocks infrared heat radiation from a planet's surface, preventing it from getting out into space. So sunlight warms Venus to an extremely high equilibrium temperature.

The atmospheric pressure and temperatures of around 460°C (860°F) were confirmed by direct measurement when the Soviet Union parachuted the first probes onto Venus in the mid-1970s. This was a difficult feat; the earliest Soviet probes were destroyed by the pressure and temperature before reaching the surface. And there was another problem. The clouds of Venus consist of sulfuric acid droplets (instead of water droplets, like our clouds) and tended to corrode the instruments on the Venus probes.

Venus is completely covered by almost featureless, yellowish-white clouds. Ultraviolet photos reveal swirling cloud patterns, but these are mostly invisible to the eye. Prior to the 1970s, therefore, there was intense speculation on the nature of Venus's surface. Suggestions ranged from dusty deserts to swamps inhabited by dinosaurs. Early observers mistakenly assumed that the clouds were made of water droplets and pictured continual, torrential rains. Soviet probes *Venera 9* and *10* transmitted surface photographs in 1975, answering some of the questions about surface conditions. They showed landscapes of rocky rubble at one site and dusty, flat rock outcrops at another. The lower atmosphere was remarkably clear. Other instruments indicated granitic and basaltic rocks and soil, similar to volcanic or igneous soils found on Earth.

The United States's *Pioneer* missions to Venus in 1978 parachuted new probes through the clouds and collected more atmospheric data (the probes were not designed to gather surface data). The clouds are concentrated in a layer at altitudes 48 to 58 kilometers (30

50

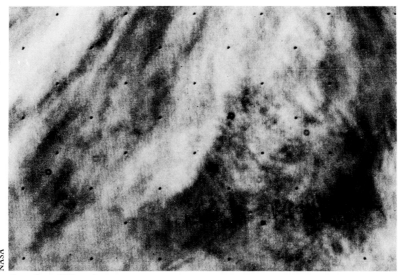

NASA

A region of Venus's equator, photographed in ultraviolet light (above).

Even nearby, Venus (left) reveals little more of itself than can be seen through a telescope back on Earth. Its thick blanket of opaque clouds (sulfuric acid gives them their creamy yellow color) is still impenetrable and nearly featureless. From this side of the clouds, it is easy to understand why Venus was named for the goddess of love. But the brilliantly pearly globe (so bright that it is capable of casting shadows back on Earth) belies the hellish inferno at its surface.

Faint, cloud patterns can be seen in the lit crescent. Venus rotates so slowly (only once every 243 Earth days) that its sunward side becomes overheated, sending its atmosphere racing at over 350 kilometers per hour (217 mph), three times faster than hurricane force, toward the cooler night side. The winds travel only in one direction—the same direction Venus spins—perpetually westward. The uppermost winds circle the planet in only four Earth days.

The night side of Venus glows faintly, like "the old moon in the new moon's arms." This dim phosphorescence has been seen from Earth for many years and is still mysterious. It may be due to a combination of causes: chemical reactions in the atmosphere, some electrical phenomenon related to the perpetual lightning deep within the clouds; or it may be similar to an Earthly aurora.

to 36 miles), much higher than normal terrestrial clouds. The particles in the clouds have a cyclical life history. Sulfur compounds condense into tiny crystals in a layer near the tops of the clouds. They react to form sulfuric acid droplets, which begin to fall as they grow. However, updrafts tend to catch them and carry them back up through the clouds, just as raindrops in our own thunderclouds may cycle up and down.

Eventually, they get large enough to fall out of the lower surface of the clouds as a "rain" of sulfuric acid, actually detected by *Pioneer* probes just below the clouds at an altitude of around 31 to 38 kilometers (19 to 24 miles). Particles detected by *Pioneer* were microscopic, but larger particles can also form.

But the "rain" on Venus never reaches the surface. The temperature increases rapidly from a "pleasant" 13.3°C (56°F) at the cloud tops to 220°C (428°F) at twenty-five kilometers below the clouds, at an altitude of thirty-one kilometers. Therefore, the droplets evaporate as they fall out of the clouds. None survive below about thirty-one kilometers, and *Pioneer* and Russian probes observed open air and a surprisingly clear landscape below this level. The light level below the clouds is something like that of an overcast day on Earth.

We're looking over the rim of a smoldering cinder cone at the towering walls of Diana Chasma (overleaf), a rift valley canyon that may dwarf even the Valles Marineris of Mars. Lightning explodes from sulfuric acid-laden clouds above a surface that is heated to a searing 460°C (860°F).

Most days are calm, which is fortunate for us, because Venus's atmosphere is extraordinarily dense. Carbon dioxide is a heavy gas to begin with, and at the crushing pressure at ground level (more than ninety times that at sea level on Earth), moving through it is like pushing through some sluggish liquid. Trying to stand in even a slight breeze, only 3 or 4 kilometers per hour (1.8 to 2.5 mph), would be like fighting the current of a river —we would be swept away as if we were in the searing exhaust of a blast furnace.

The corrosive atmosphere and high temperatures combine to crumble the landscape around us. Objects more than a few hundred meters away shimmer in the intense heat. There is a perpetual tympani roll of thunder, like the sound at the brink of an enormous waterfall. When we walk, it's in the sluggish fashion of deep-sea divers; and if you were to drop a flat object, it would flutter in slow motion like a coin dropped in water.

Ron Miller

Diana Chasma

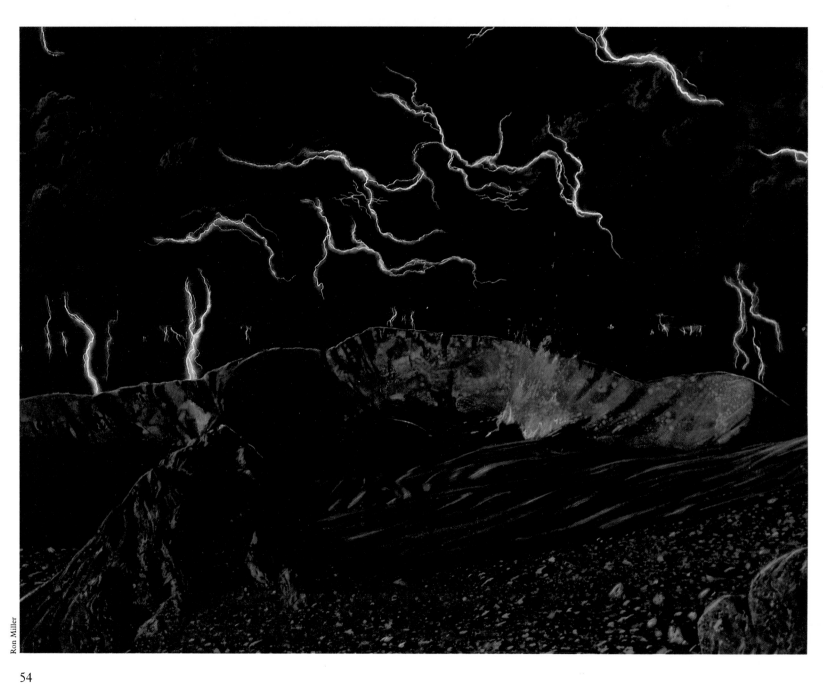

54

Ultraviolet photography enhances the contrast of Venus's radiating cloud belts.

We're standing on the rim of a volcanic caldera (left), as close as we dare to come to an active lava fountain. The landscape around us is pitted and corroded by the intense heat and corrosive rain. In the distance, roiling billows of sulfur-laden clouds are laced with lightning. Venus provides an almost continuous display of electrical fireworks: frequent flashes in the clouds provide a flickering glow while we're engulfed by the never-ending roll of thunder.

Taken in ultraviolet light to emphasize cloud structure, this photo (left) shows the weather patterns caused by clouds rushing westward at 100 kilometers per hour from Venus's sunlit hemisphere.

The *Pioneer* orbiter contained equipment that beamed radar down through the clouds and mapped the surface topography. The results suggest that Venus obeys our rule of thumb— geologic activity decreases with decreasing planet size. Venus is slightly smaller and generally smoother than Earth: Its surface consists mostly of gently rolling plains and subdued circular features that scientists assume are partly obliterated impact craters. This obliteration could have been caused by several factors, such as the filling of craters by windblown dust. Alternatively, the subsurface rocks may be so hot that they flow slowly during geologic time, flattening craters in the way that holes excavated in hot tar tend to flatten.

More interesting are a few mountainous highlands about the size of Australia, which seem to be proto-continents. These average 3 to 5 kilometers (10,000 to 16,000 feet) above the plains, but have mountain peaks rising as much as 10 to 15 kilometers (33,000 to 49,000 feet). The highest peaks exceed the altitude of Mt. Everest above our own sea floors, and are possibly volcanic. Consistent with this, the Russian *Venera* probes found basaltic lavalike soil compositions at either side of the flanks of one mountainous area.

On the other hand, a *Venera* measurement in another region suggested granitic rocks, like those forming Earth's continents and unlike those on the moon and Mars. This suggests that Venus has evolved more toward active continental geology than those smaller worlds. A few canyonlike features resemble Earth's rift valleys, where continental drift has split our lithospheric (solid rock) crust.

But the general flatness of Venus's plains indicates that Venus has not developed Earth's full-fledged plate tectonics—that is, continental drift has not acted to tear and crumple its surface to the extent found on earth.

Venus provides a vivid indication of the difficulties involved in the creation and maintenance of a habitable planet. Here is an Earth-sized globe only 30 percent closer to the sun than Earth is; yet it has a climate that could scarcely be more different. Some theorists have calculated that if Earth was only 10 percent closer to the sun, the combination of increased solar heat and the greenhouse effect would have boiled off much of our oceans, leaving more water vapor and carbon dioxide in the air. This in turn would have led to an even stronger greenhouse effect and more warming. Thus, a slight decrease in solar distance (or increase in the sun's radiation output) could have disastrously altered the climate of Earth. A small change in environmental "input" can be magnified through environmental feedback into a large change in final climatic "output." Venus's searing heat, thick carbon dioxide atmosphere, fearsome acid clouds, and thunderous

55

500 km

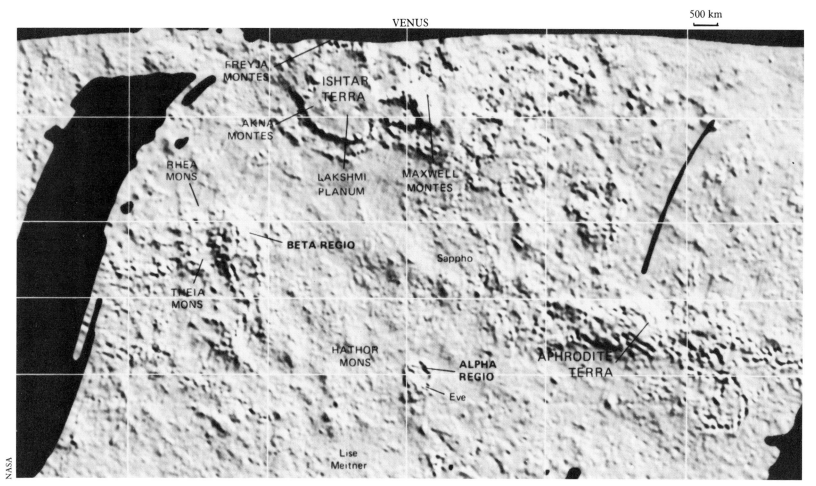

FREYJA MONTES

ISHTAR TERRA

AKNA MONTES

RHEA MONS

LAKSHMI PLANUM

MAXWELL MONTES

BETA REGIO

Sappho

THEIA MONS

HATHOR MONS

ALPHA REGIO

APHRODITE TERRA

Eve

Lise Meitner

NASA

lightning blasts make us stop and think about the consequences of our own alterations of Earth's atmosphere, which have already increased its carbon dioxide content and produced acid rains in industrialized regions.

Crossing the rubble-strewn plains not far from the north pole of Venus, we reach the rugged foothills of Ishtar Terra (left). Ishtar is an enormous block of Venus's crust, a vast plateau the size of the continental United States. In the eastern part of Ishtar Terra—invisible behind the sulfurous haze and rippling heat distortion—is Maxwell Montes, an isolated mountain complex higher than Mt. Everest.

Fumaroles smolder through mineral-encrusted vents near us. And no wonder; the dense air around us is heated to a scorching 460°C (860°F), and even higher. It is hot enough to melt lead, zinc, and tin, and to vaporize many compounds.

Ishtar was named after the Babylonian goddess of love, just as the planet itself was named for the Roman goddess. Most of the features on Venus have been named for women: a plateau named Aphrodite, a caldera called Eve,

mountains named Colette, Sacajawea, and Rhea. Yet conditions on the planet of love comes close to duplicating our idea of Hell.

56

MARS THE RUSTED PLANET

Average distance from the sun:
227,392,000 km.
Length of year: 687 days
Length of day: 24 hours 37 minutes
Diameter: 6,794 km.
Surface gravity (Earth = 1): .38
Composition: iron, silicates; carbon
dioxide atmosphere

For many years Mars has gone unchallenged as *the* planet of mystery. It is about half the size of Earth. As early as 1800, telescopes began to show polar ice caps and clouds shifting against a reddish landscape. Astronomers with improved telescopes began to report patchy and streaky Martian markings that often darkened during the Martian spring.

Around 1895, the colorful American astronomer Percival Lowell put all of this together into a magnificent new theory. Mars, he said, was populated. Not only had life forms evolved into a civilized state; the planet itself had evolved. Water vapor and other gases had slowly leaked off into space because of Mars's slight gravity, so that the air was thin and dry. The streaky markings, which Lowell drew as straight lines, were canals or vegetated land alongside of canals. Martians constructed the canals to carry water from the polar ice caps to those dry equatorial regions that were warm enough to live in. The seasonally changing markings were vast fields of vegetation.

Lowell's theory was wonderful, but it was all wrong. Even in the 1920s, astronomers accumulated evidence that the Martian air was even thinner and dryer than Lowell thought, and that his straight canals were only streaky alignments of patchy markings.

The pendulum of Martian theory swung the other way in 1965 when *Mariner 4* returned the first close-up pictures, showing not ruined cities but moonlike craters. No canals existed. Many analysts jumped to the conclusion that Mars is a moonlike world, geologically and biologically dead.

Not so. The 1971 flight of *Mariner 9* revealed an astounding discovery. Mars, a planet with no known liquid water, has winding valleys that appear to be dry riverbeds. They are called channels (although they have nothing to do with the nonexistent canals). They travel downhill, are joined by tributaries, show streamlined deposits on their floors, and empty into broad plains, sometimes with delta deposits at their mouths. In short, they have all the properties of riverbeds. How could the desert planet Mars have had abundant rivers in the past? How *old* are the channels?

One clue lies in the inventory of water on Mars. Water is hidden in three places: frozen in the polar ice caps, chemically bonded in the minerals of the soil, and frozen underground in permafrost layers like those in our arctic tundra. One possible mode of channel formation apparently involves local underground heating by geothermal activity associated with the huge Martian volcanoes (also discovered by *Mariner 9*). Geothermal heating created localized "Martian Yellowstone Parks," where vast catastrophic floods of water gushed forth, eroding channels many kilometers wide and hundreds of kilometers long. Evidence for this process includes channels that start in jumbled box canyons called *chaotic terrain*. These depressed areas of crumbled

Like an avalanche, a mountainous wall of yellow dust—hundreds of meters high—bears down upon a typically Martian landscape (right). Distant eroded hills show the effects of endless dust and wind. Once the storm reaches us, we could remain blanketed within a dull, ochre pall—through which the sun barely glows—for weeks, while winds shriek around us.

MARS

Fog fills the canyons in a view (below) *of a portion of Valles Marineris.*

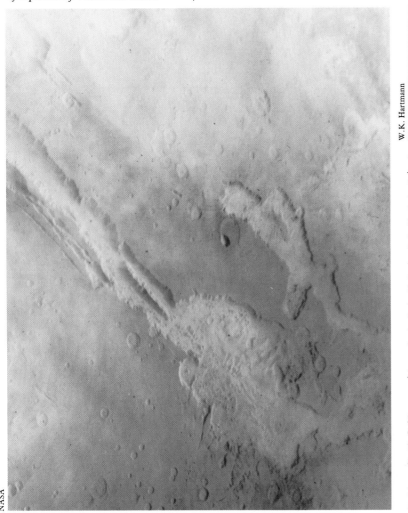

As early morning mists burn off, the fault-cracked floor of a typical branch of the Valles Marineris, named for the Mariner 9 spacecraft that discovered it, is revealed (above). *Mesas a hundred meters and more in height look like the stumps of petrified trees beneath canyon walls more than ten times higher. Valles Marineris is not a water-carved gorge like the Grand Canyon. Rather, it is the result of enormous blocks of Mars's crust moving apart—the opposite walls of the canyon were once touching. This is just how Earth's Atlantic Ocean formed, and like it, Valles Marineris has a "mid-Atlantic Ridge" running down its center.*

ground at the heads of some channels apparently collapsed as the underground ice disappeared and water gushed forth.

A second process of channel formation is called *ground-water sapping*, and is known in arid regions of Earth. If a landslide exposes an underground permafrost layer, the water can melt (or sublime directly from ice to vapor), causing the cliff face to cave in and eat its way backward into the landscape by successive collapses. This type of channel could grow from its mouth through plateau country toward the apparent "headwater," even though the latter was not the source of the water.

Do these two processes account for all Mars channels? Perhaps not. The Martian air has a pressure that ranges from only around three millibars in the high areas to as much as ten to fifteen millibars in the lowlands, compared to a thousand millibars on Earth. In the regions with pressure less than six millibars, liquid water spontaneously boils away, and even in other areas it

Ron Miller

A giant Martian "smiley face" dominates this orbital view of Argyre Planitia (below). Layers of haze, 25 to 40 kilometers high, hover over the horizon, 19,000 kilometers away.

NASA

Mars (left) seen from its smaller, outer satellite, Deimos, a dark lump of rock barely fifteen kilometers (ten miles) wide. The dusky markings on Mars's ruddy surface are those that for decades made astronomers hope—in vain—that at least plant life existed on the planet.

61

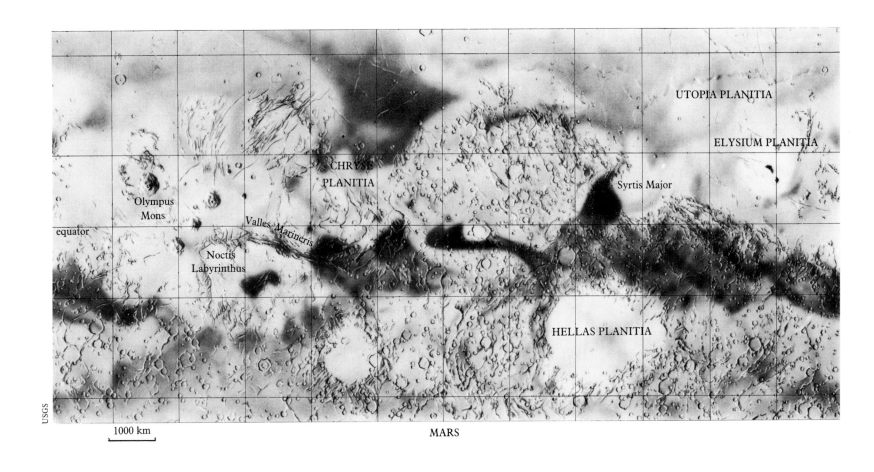

UTOPIA PLANITIA

ELYSIUM PLANITIA

CHRYSE
PLANITIA

Syrtis Major

Olympus
Mons

equator

Valles Marineris

Noctis
Labyrinthus

HELLAS PLANITIA

1000 km

MARS

Olympus Mons

Small white triangle is Mt. Everest to the same scale

A shrunken sun sets over the vast polar dune fields of Mars (right). Orbital photos show that the polar ice caps of Mars are surrounded by the largest dune field known in the solar system. Apparently, Martian winds transport dust into the polar regions, leaving it in dune formations of stratified deposits. Closer to the pole, the stratified deposits form the terraced hillsides that can be seen on the horizon. Erosion of these hills probably forms buttes similar to those in the American Southwest, as seen on the horizon at left. They would make an interesting target for geologic exploration, since they may contain exposures of strata laid down billions of years ago, giving clues to Mars's environment at that time. A light frosting of mixed ice crystals and dust has precipitated on the ground and sunlight glints off the frozen surface. The sun itself is a disk only two-thirds as big as it appears from Earth. Angle of view is about 55 degrees.

W.K. Hartmann

evaporates very fast. So it is hard for water to flow anywhere on Mars under present conditions. A channel might possibly extend its erosive lifetime by forming a frozen ice layer on its surface, shielding the water underneath from rapid evaporation. Nonetheless, many scientists think that the abundant Martian channels may have formed under different environmental conditions, when air pressure was higher and there was more water available. The number of impact craters on channel surfaces suggests they formed anywhere

from 100 million to 3 billion years ago. Thus, Mars may have evolved much in the way Lowell suggested, slowly losing air and drying out. Perhaps the climate 2 billion years ago was more moist and clement.

This hypothesis again raised hopes that Mars might yet be found to have supported life in the ancient pools of water that once existed on the red planet.

Mariner 9 revealed additional evidence that Mars is a dynamic planet, with considerable geologic activity. Photos showed that one side of Mars is dominated by a huge volcanic complex called the Tharsis Montes, containing lava plains estimated to be 1 to 3 billion years old, created by several enormous volcanoes. The

mightiest is Olympus Mons, towering about twenty-four kilometers (fifteen miles), three times as high as Everest's rise above sea level. Its base would cover Missouri. Olympus Mons is probably the largest volcano in the solar system. It has very young-looking lava flow slopes and a summit caldera. The paucity of impact craters on these areas suggests Olympus Mons formed within the last billion years, and that it may still be active today.

On the morning of July 20, 1976—seven years to the day after *Apollo 11* astronauts first stepped onto the moon—the *Viking 1* lander parachuted through Martian skies above the dusty, boulder-strewn plain of Chryse, turned on its braking

An unmoving sea of rust-red waves, the enormous dune fields of Mars stretch to the horizon and beyond (overleaf). We're standing in a very typical Martian landscape. Dark eroded rocks protrude from wind-rippled russet dust —dust that blankets most of the planet and turns the sky itself pink. Winds reaching speeds of up to 200 kilometers per hour (124 mph) keep the dust churning; enormous dust devils, tornadoes that whip the soil high into the atmosphere, feed monstrous dust storms that can cover the entire planet. In the far distance, at the right, a dust devil has begun its work as advance agent, and at our far left are the first signs of the coming dust storm. When the storm does arrive, it may last for weeks, the rocks, even the ground beneath our feet trembling under the howling blast. Dust will billow as high as ten kilometers and the wan sun will seldom be visible.

Buttes and flat-topped mesas, eroded by millions of years of sandblasting, give the landscape a distinctly familiar flavor: if it weren't for the salmon sky, we could easily imagine ourselves in Arizona.

The surface of Mars

MARS

Ron Miller

66

NASA

A tissue-thin coating of water frost on the Martian surface—perhaps no more than one one-thousandth of an inch thick (above).

Rising in thirty-meter (ninety-foot) steps like a giant's staircase, the layered terrain that surrounds the Martian south pole makes the landscape look like an enormous strawberry and vanilla parfait (left). *Fine dust deposited by storms and ash from volcanic eruptions built up the layers over millions of years. Sheets of water and carbon dioxide ice are buried within the deep, insulating blankets of soil. Wind erosion and evaporation of exposed ice deposits has created hundreds of miles of sinuous laminations that, from above, look like the contour lines on topographic maps. As scouring by wind-borne dust and sand continues, and as newly-exposed ice evaporates and allows some small or large sections to collapse, the gracefully curving cliffs constantly change shape and direction, writhing around the pole in ponderous, geologic slow motion.*

rockets, and bumped onto the Martian desert. *Viking 2* touched down on a similar desert plain called Utopia on September 3. The *Viking* landers' mission was to search for life.

Disappointingly, Mars turned out to be as dead as a doornail— or more so. Even a doornail might be expected to have some organic matter on it, but the Martian soil has no complex organic molecules to an accuracy of a few parts per billion. The soil samples did reveal curious chemical reactions when exposed to nutrients, but these are probably not caused by organisms, since no organic matter was found. Instead, chemists think the soil chemistry has been altered to unusual states by exposure to solar ultraviolet light.

These findings alter our ideas about life elsewhere in the universe. Laboratory experiments show that life's building blocks—amino acids and other organic molecules— *easily* form in carbon and water-rich environments similar to those of primitive planets. Even certain types of meteorites, carbon-rich stones called *carbonaceous chondrites,* contain amino acids that formed on their parent planetary bodies and not on Earth. So scientists had come to believe that life would probably arise spontaneously on any planet where liquid water persists, and that it might even evolve and adapt to later, harsher conditions. Mars, with its seasonally melting polar ice, its occasional balmy 15°C (60°F) summer days, and its ancient riverbeds, seems a plausible place for life to have evolved. In June 1976, the majority of planetary scientists probably would have bet that microorganisms would be found on Mars.

So why did the *Viking*s not find life? The answer probably lies in one of two scenarios. Ponds of liquid water may never have existed long enough in any one place to allow living organisms to evolve. Adding to the difficulty would be the high flux of ultraviolet sunlight that

MARS

breaks down organic molecules. (On Earth, this ultraviolet light is screened by the ozone layer of our high atmosphere, so that we receive much less of this damaging radiation, even though we're closer to the sun. Still, even a short overexposure can result in a painful and possibly even fatal sunburn.) In this scenario, Mars has been permanently without life.

In the second scenario, primitive life did evolve 4 billion years ago, when it had more abundant liquid water and a thicker atmosphere. But the atmosphere thinned and Mars grew cold and dry. The organisms all died billions of years ago. For a while, organic remains enriched the soil, but erosion broke the soil into fine dust. All of this dust, over the course of eons, has been blown into the thin air and exposed to ultraviolet sunlight, destroying the ancient organic molecules. In this view, Lowell was partially right in hypothesizing that Martian life was killed by the planet's evolution—his timing was just 3 billion years off.

Regardless of the question of life, Mars remains a fascinating environment, strangely reminiscent of our own world.

NASA

The streamlined forms in this landscape in the Chryse Basin (above) *are powerful evidence of erosion by flowing liquid—probably water flooding from melting permafrost.*

NASA

A wide-angle panorama (above) *of the* Viking 1 *landing area. A dusty sky looks bright pink.*

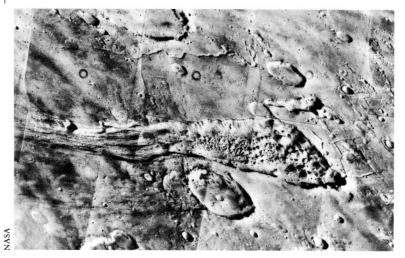

NASA

The sudden melting of underground layers of permafrost caused this area in Capri (above) *to collapse (the picture covers a region 300 by 300 kilometers). The outflowing water carved the channel at the left.*

A towering column of talcum-fine dust rears out of one of Valles Marineris's tributary canyons (right). *Throwing dust high into the thin atmosphere, whirlwinds like this feed dust to the enormous, planet-blanketing storms that can bury most of the Martian surface for days. The canyon shows the wear and tear of thousands of years of sandblasting: rounded rocks and boulders, fluted canyon walls, and mountains worn down to buttes and mesas.*

Ron Miller

70

Summer winds frequently whip up huge dust storms that may start out as similar to our dust devils, but grow to envelope much of an entire hemisphere. Air temperatures are so cold that not only water vapor freezes into puffy and wispy clouds; the main constituent of the atmosphere, carbon dioxide gas, also freezes on cold polar mornings, forming dense fogs or cloud layers of carbon dioxide mist. (The familiar "dry ice" of the ice cream vendor is this same frozen carbon dioxide.) Ice may condense on dust in the air, falling to the ground as a combination of snow and frost. The sun may burn off the carbon dioxide "snow" but leave the water on the ground. This happened during winter at the *Viking 2* site, leaving the boulder-strewn desert white with a thin layer of frost.

Much of Mars's original oxygen is locked into its soil.

Steep cliffs surround the caldera of Pavonis Mons (left), one of the smaller volcanoes that share the same plateau as Olympus Mons. Lightning flashes from the churning billows of a dust storm. The crater is many kilometers across, its floor once a flood of molten rock. No one knows how long it has been since the last eruption cascaded dust and lava over the rim of the caldera, building up the flanks of the mountain another few meters. If not still violently active, it still may sport a few hissing fumaroles, or a gas vent or geyser.

Iron-rich compounds oxidized, absorbing the free atmospheric gas, until the planet was covered with bright orange dust.

Winds have apparently transported a great deal of dust into the polar regions, forming stratified layers and the solar system's largest dune field, ringing the north polar cap. Some of these strata are eroding, leaving detached mesas and exhuming old craters that were once buried.

Cold winds blow down lonely canyon floors on Mars; morning mists form in craters and valleys; clouds condense over high volcanic mountains; Olympus Mons considers a future eruption of lava; landslides break loose and rumble down hillsides, leaving enormous scars like the scratchings of some ancient Titan; polar fogs sweep across vast vistas of dunes; the sun rises out of the dusty eastern horizon and sets in the west every twenty-four hours. All these events are happening now, even though no one is there to witness them. Mars is patiently awaiting our next visit.

VIKING
PANORAMAS

An 85-degree panorama taken by the Viking 2 lander (left). *North is to the left. It is afternoon in Utopia. On the horizon at the right are low ridges. The rocks are volcanic in origin, full of holes caused by gas bubbles, and average about half a meter in size.*

An early morning scene in Chryse (below). *It's 7:30 a.m. local time. In this 100-degree panorama (northeast is to the left) are dark, eroded rocks (the big one is one by three meters, sand dunes, and the rim of a distant crater on the middle horizon.*

MARS

Ron Miller

In some of the lowest regions of Mars, such as within the giant basin Hellas, air pressure may be high enough and temperatures warm enough to permit liquid water to exist beneath the surface (above). If you quickly dig in such an area you might find—briefly—a few drops of water glistening on your gloved hand . . . gone in a few seconds, absorbed by the dry, thin atmosphere. Still, you've witnessed something hitherto unique to Earth: liquid water in the open air. And it is not a long step from liquid water to life as we know it.

NASA

The Nile Valley (above), a lush, fertile border on either side of a watercourse— what the Martian "canals" were once thought to be.

Early morning sun floods over the rim of the greatest canyon known (right). Fog and mist still linger, made of ordinary water vapor that will disappear before the sun rises much higher. At an average lookout point, that distant sunlit cliff would be 100 kilometers away—far over the horizon. Valles Marineris invites comparison with our own Grand Canyon. Earth's great canyon stretches 450 kilometers (280 miles) through the plateaus of northern Arizona. Transferred to the United States, Valles Marineris would stretch 5,000 kilometers (3,100 miles), splitting the continent in two.

At its deepest, the floor of Valles Marineris lies five to seven kilometers (three to four miles) below us, four times the greatest depth of the Arizona canyon. The Grand Canyon would fit neatly into any one of Valles Marineris's smallest tributary canyons.

But the comparison is unfair: this great canyon is really a cousin to the Great Rift Valley of Africa—or the Atlantic Ocean. Two plates of the surface crust of both planets have split apart, and the land between has subsided. Valles Marineris may look like what the Atlantic Ocean basin was hundreds of millions of years ago. The Grand Canyon was carved by the equally tedious but much smaller-scale action of water erosion.

Valles Marineris is so long that when one end is well into night, the other end is still in sunlight. The difference in temperature between the two ends sets winds shrieking down its enormous length. Millions of years of this sandblasting and faulting have carved out the convoluted valley walls and side canyons, as well as the intricate plateaus and mesas, that seem to be floating in the sea of fog below us.

Cross-section of typical part of Valles Mari

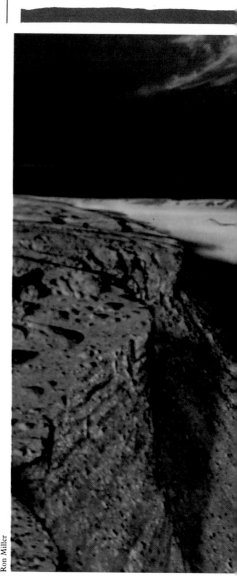

Ron Miller

Small block shows section of the Grand Canyon to the same scale

MARS

The caldera of Olympus Mons (right).
The upper crater is 25 kilometers across,
bound by sloping cliffs 2.8 kilometers
high.

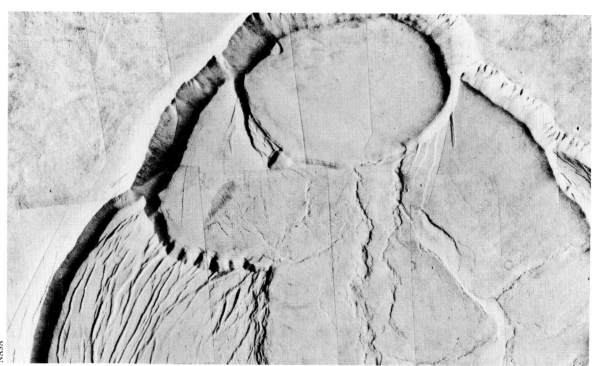

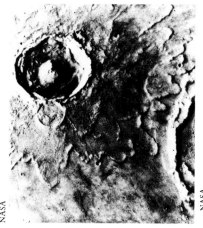

The crater Yuty (above), about eighteen
kilometers wide, made quite a splash
when it was formed. Underground ice
was melted by the impact and flowed
outward in muddy lobate flows.

An enormous landslide has bitten out a
section of Vallis Marineris's cliffs
(left). From the top of the wall to the
valley floor is about two kilometers.

Cruising several kilometers above the sixty-four-kilometer-wide caldera of the giant volcano Olympus Mons (below), we begin to appreciate the vast size of its crater. The state of Rhode Island would easily fit within its steep walls. The mountain itself is as large as Missouri and twenty-four kilometers (fifteen miles) high—more than fifteen kilometers (ten miles) higher than Mt. Everest.

We're above most of the Martian atmosphere. We have left the dust below us and the sky is turning to indigo. Water vapor crystallizes in the thin, frigid air and wave clouds form as winds pour over the broad summit.

Is Olympus Mons still active? No one knows for sure, but strange clouds observed over the years and photographed from orbit may be signs of volcanic life.

Ron Miller

GANYMEDE A NEW WRINKLE

Distance from Jupiter: 1,070,400 km.
Revolution: 7 days 3 hours
Diameter: 5,216 km.
Composition: ice, carbonaceous silicates

Ganymede, the largest satellite of Jupiter, is also the largest satellite in the entire solar system. It has a diameter of 5,270 kilometers—greater than that of the planet Mercury. From earth, it appears the size of a pinhead in large telescopes, with only shady markings visible on the clearest days. Spectra reveal that there is little or no atmosphere on Ganymede, and the surface is dominated by frozen water and some soil. At this distance from the sun, ice is stable even when exposed to full sunlight, and it is therefore usually solid. Only volcanic heat from the interior is likely to cause volcanic eruptions and melt the ice, allowing water to seep under the surface and freeze in formations resembling lava flows. The reflectivity of Ganymede is several times higher than the reflectivity of our moon's dark rocks, in keeping with Ganymede's surface mixture of bright ice and dark soils.

One of the best clues about the interior of Ganymede comes from its mean density, i.e., total mass divided by total volume: it is higher than the density of pure ice but lower than that for most rock types. This indicates that the interior is also a mixture of ice and rock. Most of the rock has probably sunk through the ice to form a rocky core surrounded by an ice mantle.

The major importance of Ganymede turned out to be its transitional position between the geologically active planets (Earth, Venus, Mars) and the small worlds that cooled too rapidly to develop geological activities such as eruptions and faulting. Like Mars, Ganymede has some regions of ancient, heavily cratered crust and other regions that have been resurfaced. The old regions are quite dark; dark soil seems to have accumulated as ice was vaporized by meteorites.

Among the densely packed craters of these regions are parallel sets of long curved furrows a few kilometers wide and hundreds of kilometers long. These seem to be fragments of concentric ring systems that surround the largest impact craters. Well-preserved craters in this class have been found on the moon and Mercury, and are known as *multi-ring basins*. Apparently, the most energetic meteorites do not excavate simple craters, but rather set up activity in the surface that results in the production of concentric, clifflike rings. Complete systems of these concentric rings are found on neighboring moon Callisto as well as on Mercury and our own moon, but on Ganymede many areas have been resurfaced, destroying all but fragments of the multi-ring systems.

The surface of Ganymede looks like a giant jigsaw puzzle, pieces of which have been removed and replaced with some different kind of material. Certain regions of the original picture are preserved, but others have been destroyed by resurfacing. On Ganymede, the newer regions are lighter, wandering in wide bands among the darker polygonal cratered areas. The lighter regions are traversed by intricate systems of

A tedious climb up a long slope, ice powdered by millions of years of micrometeorites crunching underfoot, has brought us to the top of one of a series of parallel ridges that wrinkle the surface of Ganymede like a washboard (right). These thousands of square miles look like a titanic Japanese sand garden. It's about ten kilometers (six miles) to the next ridge, and about a hundred meters (300 feet) to the bottom of the valley below us. Climbing the ridge has been like crossing a glacier back on Earth. Ganymede's surface— and its bulk, to a great depth—is made of simple water ice, mixed with a little dust and rocks. Its icy landscape has long ago been pulverized.

Jupiter sits 890,000 kilometers (551,800 miles) away, the Great Red Spot peeking coyly over the horizon. Above the planet are two more of the four Galilean satellites: Europa and orange Io. The sun is at our backs and Jupiter is directly in front of us, so we can see the shadow Ganymede casts, the small black spot near the center of Jupiter. Since the shadow is nearly the same size as Ganymede itself, you can get a good idea how much larger Jupiter is than its relatively tiny moon.

Ron Miller

narrow fissures, usually arranged in parallel streaks, here and there broken by craters.

Analysts suggest that the original dark crust broke apart due to internal heating and expansion, allowing Ganymede's equivalent of lava—erupting water—to flow out onto the surface and freeze into ice flows. Subsequent eruptions along the fissures squeezed out new parallel ridges of ice. The same phenomenon also happens in our Arctic ice floes.

Ganymede may well be showing us an incipient process of plate tectonics, thus adding to our understanding of Earth. On Earth, the lithosphere—or solid rock layer—long ago broke into plates that ride on a partly molten lower layer. Dynamic currents in this lower layer, driven by Earth's internal heat, drag the surface plates, causing collisions and throwing up mountain ranges. On Ganymede,

GANYMEDE

the process is not as dynamic. Ganymede's interior may be warm, but it is not as hot as Earth's. The ice "lithosphere" at the surface may have broken into "plates," but the plates have only jostled one another. Little pieces here and there seem to have moved tens of kilometers, offsetting craters and faults, but there have been no large-scale motions or collisions and no folded mountain ranges like the ones found on Earth.

Ganymede has other interesting features. Whitish polar caps, seemingly thin water frost deposits, extend down to latitudes around 40 degrees. The poles are the coldest regions because they get the least direct

Ron Miller

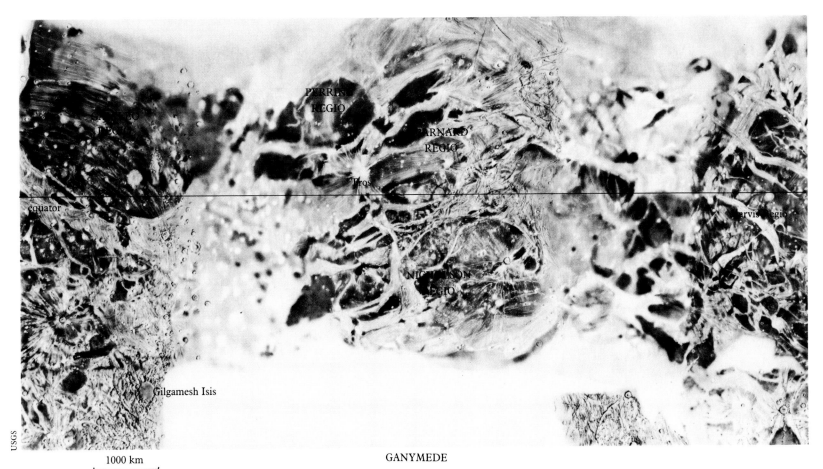

1000 km

PERRINE REGIO

BARNARD REGIO

Tros

NICHOLSON REGIO

equator

Jarvis Regio

Gilgamesh Isis

GANYMEDE

A short descent takes us into the shallow, curved valley between the crests separating Ganymede's grooves (left). Although we are standing in shadow, our surroundings are brightly lit by Jupiter and by sunlight reflected off the sloping wall of ice opposite us.

sunlight, and it may be that water vapor escaping from fractures around the planet condenses in frost fields there.

Most fresh craters have bright radiating rays of ejecta blown out onto the surrounding surface. These craters have probably blown away the darkish surface soil, exposing fresher ice. But occasional small craters, typically ten to twenty kilometers (six to twelve miles) across and a few kilometers deep, have dark rays. They probably reflect a layered structure beneath Ganymede's surface. If surface layers ever melted or if "volcanic water" gushed out carrying soil with it, the soil would sink before the resulting oceans froze. Thus, craters penetrating to a few kilometers may tap layers of blackish soil. Ganymede's surface may contain intricate sedimentary layers of ice and rock.

Ganymede provides us with a simple model in which ice, one of the prime geologic constituents, interacts with rocky material. The remnants of cratered "plates," partially erased by fractures and ice flows, testify to the forces that shape all worlds.

GANYMEDE

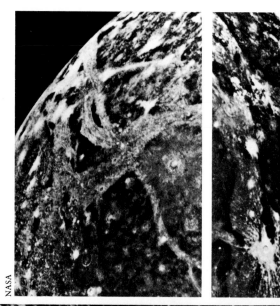

NASA

Two close-up views of piebald Ganymede taken by Voyager 2 *(left). Both show different aspects of the block of dark, heavily cratered terrain that distinguishes Ganymede. The light parallel lines may be related to an ancient impact, similar to the one on Callisto. Recent craters are surrounded by halos of fresh ice.*

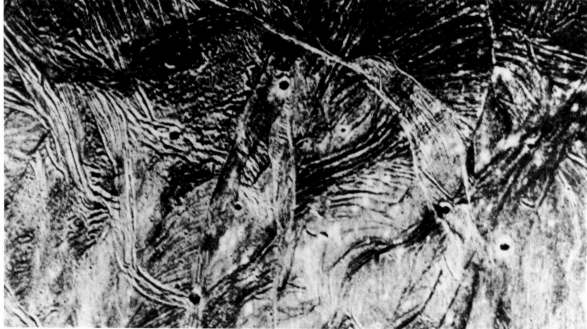

NASA

Complex patterns of ridges and grooves (left) swirl in an area of Ganymede's surface about 580 kilometers across (the smallest feature visible is about 3 kilometers wide). Younger grooves are superimposed on older ones.

TITAN THE MASKED PLANET

Distance from Saturn: 1,222,000 km.
Revolution: 16 days
Diameter: 5,500 km. (?)
Composition: (?); hydrogen atmosphere

In 1944, the Dutch-American astronomer G.P. Kuiper wrote a paper with the surprising title, "Titan: A Satellite with an Atmosphere." This title was surprising because until that time, all satellites had been assumed to be desolate, airless worlds something like our own moon. But Kuiper's spectra showed strong absorptions of certain colors due to methane gas on Titan.

Artists such as Chesley Bonestell quickly recognized the potential beauty of Titan; a gaseous atmosphere of methane would scatter sunlight with a sky-blue color similar to that on Earth. Bonestell produced stunning paintings of a yellowish-tan Saturn hanging in the beautiful blue sky of Titan over rugged mountains and white snow fields. Unfortunately, observations in the early 1970s revealed the light we see from Titan is reflecting off clouds or a thick layer of haze. This means, to the dismay of space artists, that Titan's surface probably is completely obscured by clouds, making Saturn invisible to an observer standing on Titan . . . and making the scene in Bonestell's painting impossible.

Meanwhile, astronomers debated the composition of Titan's atmosphere. In some models, methane was the major constituent and the atmosphere was very thin. Other models suggested that nitrogen, probably released from Titan's interior, might form a major portion of the atmosphere, perhaps with small amounts of hydrogen mixed in as well. Chemical models suggested that chemical compounds formed from the reaction of these materials would give the clouds their observed orange-red color. Some theorists pointed out that very large amounts of nitrogen might be present in Titan's atmosphere without being observable from Earth, so that instead of a wispy, thin atmosphere, Titan might have an atmosphere as dense as Earth's.

In 1980, *Voyager 1* flew through Saturn's system of satellites and was programmed to pass close to Titan. Scientists hoped that its cameras would not only photograph the clouds, but would also penetrate the haze and reveal the nature of surface features. Would they see a cratered landscape with little geologic activity, or a terrifically

Ron Miller

In our descent toward the surface of Titan (above), we pause for a minute a few score kilometers above the sluggish orange clouds to enjoy the sight of a crescent Saturn, its arc bisected by the thin line of the rings, like a drawn bow and arrow suspended in a hazy blue sky. The turgid clouds beneath us are almost opaque, barely allowing us to see the dim glow of Saturn from Titan's surface.

84

active landscape with geysers of water erupting through clouds of methane and steam? Perhaps the haze would be thick enough to obscure nearly all the surface, with only an occasional volcanic mountaintop protruding above the clouds.

The *Voyager* cameras actually revealed a world that was disappointing in terms of atmospheric clarity, but terribly exciting as an example of planetary evolution. Titan appeared as an orange ball with no surface features visible and with only a few mottlings in the atmospheric cloud layer. A dark-toned polar cap was visible, and the cloud surface seemed to be brighter in the southern hemisphere.

There is some suggestion that bands or contrast shadings run parallel to Titan's equator. But the *Voyager* instruments strongly

confirmed the thick-atmosphere model of Titan. The spectra of nitrogen recorded by these instruments indicated that nitrogen is much more prevalent than methane. This means that Titan corresponds to the most dense-atmosphere models proposed before *Voyager*. *Voyager* instruments measured pressures at Titan's surface of roughly 1.6 times the air pressure at sea level on Earth.

Many scientists had hoped that if the air of Titan turned out to be that thick, it might produce a warming or blanketing effect similar to the greenhouse effect on Venus and Earth. If this were true, Titan might be considerably warmer than predicted, considering its distance from the sun. However, preliminary measurements from *Voyager* instruments suggest that most of Titan's atmosphere is very cold, with typical surface temperatures around $-180°C$ ($-292°F$).

Nonetheless, the *Voyager* discoveries indicated a phenomenal world. *Voyager* instruments revealed not only methane and nitrogen, but other gases such as ethane, acetylene, ethylene, and hydrogen cyanide. *Voyager* scientist Don Hunten commented that ultraviolet radiation from the sun "would

work on the methane in the upper atmosphere of Titan to produce octane, which is, after all, the main component of gasoline . . . I can imagine it raining frozen gasoline on Titan." Indeed, chemical studies suggest that many of the heavier compounds formed by reactions in its atmosphere would precipitate onto Titan as rain, perhaps intermittently.

In the environmental conditions of the surface, methane may exist as snow, liquid *and* gas, and some nitrogen might be in liquid form. Titan may have large expanses of liquid methane or nitrogen ocean. *Voyager* scientist Dr. Von R. Eshelman characterized Titan's surface as "A bizarre, murky swamp. . . . The swamp is liquid nitrogen and the murk is frozen nitrogen and hydrocarbon muck." Since Earth's atmosphere is 80 percent nitrogen, the conditions that seem to exist on Titan have some resemblance to early chemical conditions postulated for Earth. For this reason, Titan may shed some light on organic chemical reactions in bizarre planetary environments . . . and therefore, on conditions that lead to life.

In any case, *Voyager* has revealed an extraordinary new

world that teasingly provokes our imaginations. Perhaps there are places where the clouds are thin and the cloud-filtered glow of Saturn and the remote sun illuminate rivulets of liquid methane or organic slime that run down eroded hillsides into vaporous, frigid swamps. Perhaps eruptions produce volcanoes of ice, the red clouds above occasionally parting to reveal the beautiful sight of Saturn hanging in a blue sky.

A wan sun glows through jets of liquid nitrogen and methane as a geyser erupts through a slushy basin on Titan's dim landscape (left). *Internal heat may provide the energy for these graceful—if frigid—fountains. A rare and fleeting clearing in Titan's perpetual orange haze gives us a brief glimpse of distant crescent Saturn.*

Pools of liquid nitrogen mirror the hazy image of Saturn, seen through a murky veil of organic compounds (overleaf). *Titan's surface is coated with these orange and brown compounds that perpetually sift down from the clouds. Patches of frozen methane relieve some of the monochromatic monotony.*

Titan

CALLISTO THE BATTERED PLANET

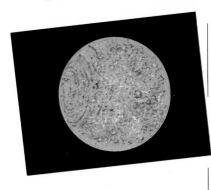

Distance from Jupiter: 1,882,600 km.
Revolution: 16 days 16 hours
Diameter: 4,890 km.
Composition: ice, carbonaceous silicates

With a diameter of 5,000 kilometers (3,100 miles), Callisto is the small sister of Ganymede. It seems to have lacked the internal energy to disrupt its own surface, and it resembles what Ganymede might have looked like in an earlier stage of its evolution. Callisto's surface is virtually totally cratered and dark, and has not been broken into "plates" like the surface of Ganymede. Here and there huge bull's-eyes of concentric rings seem to mark the sites of the largest ancient impacts.

The bulk density of Callisto is surprisingly low, indicating that Callisto is mostly ice, with only a small mixture of rocky material. Spectroscopic observations from Earth indicate that it is covered by a dark, rocky dust layer, possibly similar to the blackish carbonaceous meteorites that are believed to come from this part of the solar system and from the outer asteroid belt. This soil layer may accumulate as impacts of small meteorites vaporize the ice and leave black, carbonaceous material behind.

The multi-ring systems on Callisto are virtually flat; its bigger craters are shallow. The reason for this probably involves Callisto's icy composition. Because these craters are formed of ice, and because ice flows (as glaciers do), the deepest craters have leveled out. This means relief on Callisto is rarely more than a kilometer. The biggest impact sites, originally marked by deep basins and surrounding vaulted cliffs, have filled out to the point where only ghostly ring scars remain. Small craters

An enormous bull's eye (above), *created by the impact of a small asteroid. Waves of molten surface ice spread for hundreds of kilometers.*

preserve their nearly original profiles because they aren't big enough to deform. To put it another way, the ice is rigid enough to hold up small crater rims, but not large crater rims.

Callisto seems to be the best example of a world not quite big enough to be deformed by internal energy, and not quite close enough to another planet to be repeatedly stretched and heated by tidal forces. Callisto is one of the largest worlds that has survived from the planet-forming era with little internal modification.

NASA

87

CALLISTO

Callisto's surface (below) is saturated with craters; Callisto literally could not be more cratered. New craters could not be formed without destroying old ones. Far from the crust-twisting influences of Jupiter's gravity,

Callisto's icy surface froze early; with the exception of the impact of the occasional meteorite, little has happened on Callisto since it was formed.

Ron Miller

Ron Miller

Dirty ice and bare rock surround us (above), mixed into a muddy slurry by one of the enormous impacts that scarred Callisto's surface and then, almost instantly, froze.

Jupiter, bisected by its faint ring, accompanies Europa, Ganymede, and Io in Callisto's dark sky.

CALLISTO

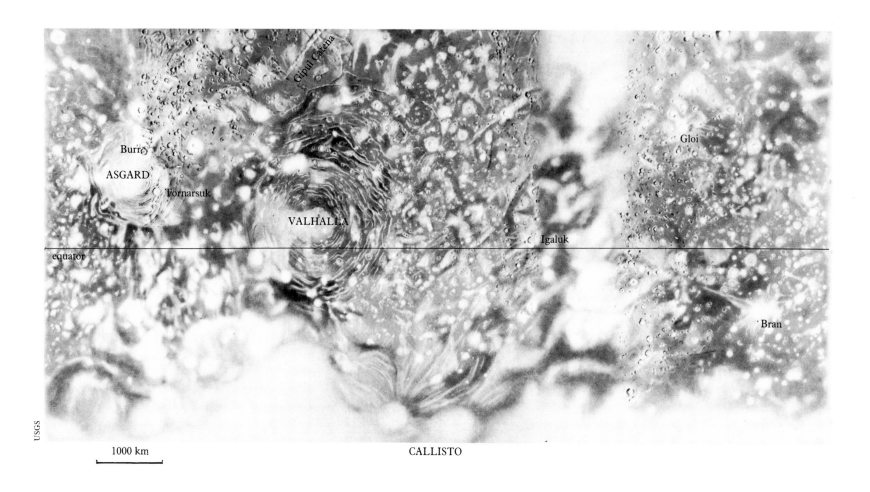

ASGARD
Burr
Tornarsuk
Gipul Catena
VALHALLA
Igaluk
Gloi
Bran

equator

USGS

1000 km

CALLISTO

IO A WORLD TURNING INSIDE OUT

Distance from Jupiter: 421,800 km.
Revolution: 1 day 19 hours
Diameter: 3,636 km.
Composition: silicates, sulfur

To those of us who grew up walking on a surface of silicate soil, watching trees sway in the breeze under a blue sky, dangling our toes in liquid water pools, throwing snowballs, and watching astronauts play in the rocky deserts of our sister world, the moon, Jupiter's satellite Io must appear to be the most bizarre world in the solar system. Io is almost exactly the same size as our moon, but its geology and chemistry are dominated neither by the silicate soils of the inner planets nor by the ices of the moons of the giant planets, but rather by volcanic, sulfur-rich compounds.

The first close-up pictures by *Voyagers 1* and *2* in 1979 showed Io to be mottled with orange, yellow, red, and white patches, and pocked with blackish spots. *Voyager* team scientists remarked that they didn't know what was wrong with Io, but it looked as though it might be helped by a shot of penicillin; Io resembled one of those planets we used to laugh at when it appeared outside spaceship windows in grade-C science fiction movies.

Within a few days of *Voyager 1*'s approach, Jet Propulsion Lab navigator Linda Morabito was studying the outline of Io when she discovered a huge, faint plume of diffuse material extending out from the surface. A total of eight such plumes were soon identified, shooting 70 to 280 kilometers (43 to 174 miles) above the surface, and spreading to widths as great as 1,000 kilometers. They were erupting out of violently active volcanoes, at velocities of one-half to one kilometer per second (1,100 to 2,200 mph).

Four months later, when *Voyager 2* flew by Io, at least six of these volcanoes were still active, and two new ones had started. Close-up photos revealed that most of the blackish spots were volcanic calderas, the irregularly shaped craters left by volcanic eruptions. Io maintains continuous simultaneous major eruptions—a level of volcanic activity unheard of on Earth, where violent eruptions occur only sporadically.

While the normal surface of Io has a daytime temperature of about −148°C (−234°F), numerous warm spots have temperatures about 27°C (81°F), while frequent eruptions have temperatures around 327°C (621°F). *Voyager* data suggest outbursts as hot as 427°C (801°F). These outbursts could thus involve molten sulfur (melting point 112°C (234°F), but would be a few hundred degrees cooler than the temperature of familiar molten silicate lava on Earth.

The intensity of Io's volcanism is emphasized by the fact that on most planets, volcanic activity has dwindled to nothing. Ancient volcanic outpourings built the lunar lava plains and the 24-kilometer (79,000-foot) Martian volcanoes—the largest in the solar system—but these heat sources have apparently mostly died out. Radioactivity produced the heat to drive volcanism on these worlds. The smaller worlds

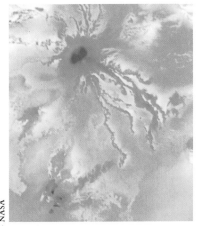

A volcanic caldera (above) *with radiating flows of lava.*

have radiated their heat away and cooled off.

What creates the energy and heat to drive the continuous volcanism of this modest world? Io's orbital position provides the answer. Any satellite close to a planet acquires a certain "stretch" as a result of that nearby planet's gravitational force. If the satellite is in an elliptical orbit, the distance to the planet and hence the stretching tidal force change, causing a flexing of the satellite. Flexing causes friction; friction creates heat. The satellite heats up as a result of this flexing. Although Io's orbit is virtually circular,

IO

Ron Miller

Debris is being spewed at a thousand kilometers per hour at us as we hover a few score kilometers above one of Io's violently active volcanoes (above). The entire visible surface of Io has been buried to an unknown depth beneath the ejecta from its almost continually erupting volcanoes. Molten sulfur has poured in streams from wounds in the dome's side, while blue gases fume from a caldera at the summit, filled with black, molten sulfur.

which seems to preclude tidal flexing as a heat source, scientists have found that neighboring satellites exert forces that make Io's orbit change slowly, causing strong flexing and the subsequent heating of Io's interior.

The volcanism of Io explains a fact that had puzzled pre-*Voyager* scientists: the absence of frozen or liquid water on its surface, discovered through observation of Io's spectrum. Jupiter's moons would be expected to have formed from a mixture of frozen water and rocky material, with minor amounts of sulfur and other compounds. Apparently, Io's volcanism has been so intense that most of the crustal materials have melted, erupted, been covered by later eruptive debris, and perhaps been recycled through this sequence again and again. Volatile materials, such as water, boiled off long ago. This is why Io lacks the ice-rich surface of its neighbor worlds, such as Europa.

Heavy materials, such as silicate rock, sank into the lower layers of Io. The retention of the heavy material and "boil-off" of light volatiles accounts for the fact that Io's density is higher than the other moons' densities. Sulfur compounds, the lightest materials that would not boil away, ended up as the dominant "lavas" on the surface.

Io has some of the most stunning visual effects in the solar system. Imagine a full

"day" on Io, experienced from a single point on Io's surface. One side of Io always faces Jupiter. On that side, Jupiter dominates the sky, always hanging in the same place. Jupiter subtends an angle of about 20 degrees—forty times the angular size of the moon in our sky.

Because Jupiter covers a large part of Io's sky, the sun spends nearly two and a half hours of each day in total eclipse behind Jupiter. When the eclipse ends, the sun comes out from behind Jupiter and begins to warm the landscape, which has cooled markedly during the eclipse. At this moment, and, from Earth, another of Io's unusual phenomena occasionally becomes visible: the so-called post-eclipse brightening.

In the mid-1960s astronomers monitored the brightness of Io before and after these eclipses, hoping to discover whether Io had an atmosphere. They hoped that during the cold eclipse period, frost might condense and

Without an atmosphere to suspend gases and dust in the cauliflowerlike billows familiar to terrestrial volcano-watchers, volcanoes on Io (right) look like vast garden sprinklers, blowing debris in graceful arcs for hundreds of kilometers. Standing inside the caldera of an active volcano, near chunks of raw sulfur blown from previous eruptions, we watch the enormous plumes as they thunder from the vent with explosive violence.

Ron Miller

IO

last for a few minutes after the eclipse, causing Io to be brighter than usual for those few minutes. Sure enough, they found that Io was often brighter than usual for ten or fifteen minutes after an eclipse. But as other observers checked this result, it became clear that it did not happen after every eclipse—only occasionally.

When later data showed virtually no atmosphere on Io, this phenomenon was put on the back burner reserved for unsolved scientific curiosities. *Voyager*, however, may have provided an explanation. One of the volcanic materials on Io is believed to be sulfur dioxide, which can condense into a

1

2

A DAY ON IO

Io photographed by Voyager 1 (above). *The feature at the right is an erupting volcano.*

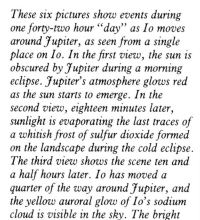

4

These six pictures show events during one forty-two hour "day" as Io moves around Jupiter, as seen from a single place on Io. In the first view, the sun is obscured by Jupiter during a morning eclipse. Jupiter's atmosphere glows red as the sun starts to emerge. In the second view, eighteen minutes later, sunlight is evaporating the last traces of a whitish frost of sulfur dioxide formed on the landscape during the cold eclipse. The third view shows the scene ten and a half hours later. Io has moved a quarter of the way around Jupiter, and the yellow auroral glow of Io's sodium cloud is visible in the sky. The bright

colors of Io's sulfurous surface are not prominent.

In the fourth view, after another ten and a half hours, the sun is setting behind us and Jupiter is fully illuminated. Shadows stretch toward the horizon. In the fifth view, ten and a half hours later, we have moved three-quarters of the way around Jupiter, and the sodium aurora is again visible in the

sky. In the sixth view, shortly before dawn, the volcano on the horizon has erupted. Delicate traceries of debris follow parabolic trajectories in Io's near vacuum instead of forming the puffy clouds of Earth's volcanoes. The highest parts of the parabola are lit by the sun, just below the horizon to the left; the lower part of the parabola is in Io's own shadow.

The plumes of Io's volcanoes squat on this moon's marbled surface (above). *The force behind these titanic eruptions blows debris hundreds of kilometers into space. Some of it spirals in toward Jupiter, eventually coating the tiny innermost satellite, Amalthea, with dull orange deposits.*

whitish frost or snow if the temperature drops low enough. Calculations suggest that if several large volcanoes are erupting during an eclipse, enough sulfur dioxide might condense on orange or red backgrounds to brighten the overall appearance of Io during the few minutes after the eclipse ends.

Some ten and a half hours later, when Io has moved a quarter of the way around

Jupiter, its sky develops a faint yellowish glow. This effect was first seen from Earth. The explanation is that a large, thin gas cloud of sodium, sulfur, and other atoms surrounds Io. The presence of this cloud has to do with Io's location inside Van Allen radiation belts around Jupiter.

Energetic atoms trapped in these belts strike Io's surface and knock other atoms loose. More may "come unglued" due to Io's

volcanic eruptions. As these atoms escape Io, they diffuse into a cloud many Io-diameters wide, stretching forward and backward along Io's orbit. Sunlight striking the atoms excites them. In particular, if sunlight of a certain wavelength strikes the sodium atoms, they absorb and then re-emit this color—a glow called the *sodium "D" line*, familiar to us as the yellow color in most candle flames. The sodium cloud around Io glows with a faint yellow light

—a sodium aurora that resembles a moderate aurora on Earth.

About ten and a half hours later, Io is halfway around its orbit, between Jupiter and the sun. The sodium glow has slowly faded from the sky. Now Jupiter is in its "full" phase—a dazzling yellow, orange, red, and tan disk with brightly colored cloud patterns, such as the famous Red Spot storm system.

The sun has set at our location after another ten and a half hours. We have now traveled three-quarters of the way around Jupiter, and the planet is in its "third quarter" phase. The yellow sodium glow is back in the sky, perhaps a bit brighter now, depending on the amount of volcanic gases emitted in the last few hours. In the cold of night, the sulfur dioxide frost may have formed again, and the landscape is dully illuminated by the yellow light of Jupiter itself.

So far, no one has stood on the plains of Io to see these views or feel the seismic tremors as its volcanoes explode. But if we humans manage to establish viable interplanetary travel before we blow ourselves away or imprison ourselves forever on a resource-exhausted planet, the day may come when, heavily shielded from Jupiter's radiation, we will see the sights of Io not through spacecraft instruments or paintings, but with our own eyes.

W.K. Hartmann

Io's eerily glowing sodium torus lights the moon's night sky (above). Sodium atoms dislodged from Io form a cloud that surrounds Io and stretches out along its orbit around Jupiter, forming a donutlike torus around Jupiter. The portions of this cloud that move toward or away from the sun as they move around Jupiter give off a faint yellow glow, the same glow that is prominent in candlelight. (We stand on a plain on the night side of Io; Io is 90 degrees from the Jupiter-sun line, and the landscape is lit by Jupiter in half phase. We are looking at the sky along Io's orbital path—a direction 90 degrees away from Jupiter, which is out of the picture to the upper right. This means we are looking down the axis of the glowing torus, and we see the brightest part of it. In the distance, the glow curves off to the right along the orbital path around Jupiter.

A red sulfur flow can be seen in the background. The neighboring outer moons, Europa (lower left) and Ganymede (upper right), hover in the sky. The visual width of this wide-angle view is about 80 degrees.

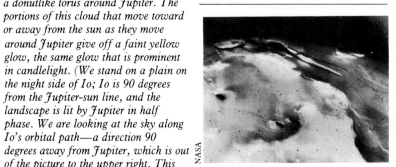

NASA

A scene along the terminator, (above) photographed by Voyager 2. The sun is shining from the lower left to the upper right. The blade-shaped valley at the top is 300 kilometers long and 50 kilometers wide.

Blue gases are venting from a caldera filled with black, molten sulfur (below).

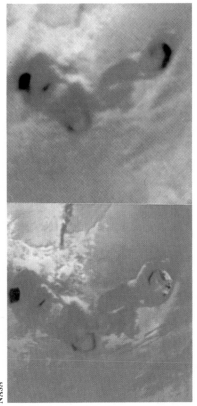

NASA

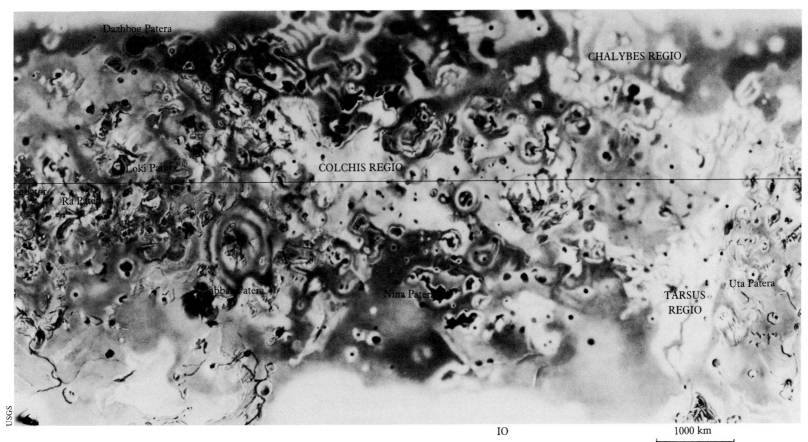

Dazhbog Patera

CHALYBES REGIO

Loki Patera

COLCHIS REGIO

equator

Ra Patera

abban Patera

Nina Patera

TARSUS REGIO

Uta Patera

USGS

IO

1000 km

Voyager *pictures showed that most of the volcanic calderas on Io* (left) *are very dark in color, probably filled with lakes of the nearly black molten sulfur that occurs at temperatures of about 227°C (441°F). In some of the* Voyager *pictures, transient blue glows can be seen in calderas. These are probably clouds of escaping gas, lit by the sun and glowing in the same way that our atmosphere does. Here we see such a cloud of gas being emitted from a slumped fracture zone along one of the caldera walls. Elsewhere, the walls reveal a striped cross-section of colored sulfur flows, similar to the strata seen in the walls of terrestrial volcanic calderas. In the distance, we see bright sulfur-yellow plains, and a blood-red volcanic cone sits ominously on the horizon. Prominent in the sky are neighboring satellites Europa* (right) *and Ganymede* (left). *The visual angle of this wide-angle view is about 65 degrees.*

MERCURY CHILD OF THE SUN

Average distance from the sun:
  57,900,000 km.
Length of year: 88 days
Length of day: 1,416 hours (59 days)
Diameter: 4,880 km.
Surface gravity (Earth = 1): .38
Composition: nickel-iron, silicates

Mercury is a tiny planet, nearly the same size as Callisto. But size is all they have in common. Instead of a frigid world made of solid ice, Mercury is a desolate cinder whose orbit averages closer to the sun than any other known world.

Mercury's closeness to the sun means that it formed from the hottest materials available in the primeval solar system. Here there were no ices, no minerals with low melting points. The dust from which Mercury aggregated consisted mostly of high-temperature minerals: metals and silicates. Supporting our conjectures is Mercury's high average density, unusually high for such a small world. Taking into account central compression by the weight of overlying material, analysts calculate that Mercury has a metal core proportionately larger relative to its size than any other planet. Mercury probably melted due to heat released by radioactive minerals, and its iron and other metals drained downward to the core.

This melting must have taken place long ago. Mercury's scorched surface has since been battered by meteorites, comets, and asteroids; it is entirely covered by impact craters. Only a few small lava plains have survived the bombardment. Geologically, Mercury appears to have been dead for 3 or 4 billion years.

The sun's tides have locked Mercury's rotation into a very slow spin: one day requires two-thirds of its year. Its rotation, relative to the stars, takes fifty-nine days, while its orbit around the sun takes eighty-eight. A day only one-third shorter than a year means that the sun moves very slowly through the sky. One hundred seventy-six Earth days must elapse between one sunrise and the next.

It is the strange things that the sun can do during one of Mercury's long days that makes the little planet worth visiting. Because Mercury's orbit is slightly eccentric (elliptical), there is a brief time when its orbital speed overtakes its rotational speed. For a little more than a week, you—like Joshua—can actually watch the sun slowly come to a complete halt. It then moves backward for a time before continuing in its original direction, having performed a complete loop in the sky. At other positions on Mercury, you can see *two* sunrises and *two* sunsets each Mercury "day."

Mercury's rugged, cinderlike surface (below).

NASA

MERCURY

By stepping into the shadow of a boulder thrown onto the rim of a small crater, we can block out the deadly glare of the sun itself, enabling us to see the delicate streamers of the solar corona, the sun's outer atmosphere (left). *This, like the sun itself, is three times closer—and three times bigger— than on Earth. Jets of ionized gas ejected from the sun are molded by the sun's powerful magnetic field on their way to Earth to disrupt radio broadcasts and power the flickering "neon lights" of the aurora borealis.*

The large, faint glow across the sky is the zodiacal light—interplanetary dust lit by the sun. In the upper right are two "stars." The brighter one is Venus and the other is Earth.

101

102

Mercury's closeness to the sun and its long afternoons make its surface very hot during the daytime, while the equally long nights get very cold. Afternoon soil temperatures can rise well above 227°C (441°F); at night, the temperature can drop to around −173°C (−279°F). The surface of Mercury closely resembles a sun-blasted, copper-colored version of our own moon. Dust is abundant, created not only by micrometeorite impacts but also from thermal erosion of rocks: chips flake off as the rocks expand and contract from the extreme day-to-night temperature changes. As old rocks erode, new ones are blasted out by fresh meteorite impacts.

Mercury, unlike our moon, has no friendly blue globe to dominate its heavens. But it makes up for that with two unique sights. In Mercury's night sky, Venus and Earth often appear as brilliant stars—bright

Mercury has two brilliant stars in its night sky: Venus and Earth. In the shadow of a hillside not long after sunset—a distant cliff is still lit by the sun's descending orb—we can look back on our home planet (left). Keen eyes may even be able to pick out the tiny pinpoint next to the brilliant blue star—Earth's moon. Far brighter, and much nearer to Mercury, is the planet Venus, only 50 million kilometers (31 million miles) away—as opposed to 91 million (56 million) for Earth.

enough to cast faint shadows. The sharp-eyed may try to see if they can find the moon, a faint pinprick next to Earth. And, with the sun so very close, Mercury gets singular sunrises and sunsets. There is no atmosphere, so there are no reds and oranges, but the sun's corona, or outer atmosphere, extends well above the horizon just before sunrise and after sunset. A dim glow extends outward from the pearly yellow corona, the faint band of *zodiacal light:* dust concentrated in the plane of the solar system and illuminated by the sun. It is sometimes visible from Earth as well.

Cutting through Mercury's pocked surface like cracks in shattered glass are several enormous cliffs. Up to 500 kilometers (310 miles) or more in length and 2 to 4 kilometers high, they pierce mountains, valleys, and craters alike. They are the result of a crumpling of the landscape—a type of feature geologists call a *thrust fault*—that happened when the planet's surface was cooling and contracting. On one side of the cliff the land has been raised, while on the other side it has been lowered. You can see huge craters that have been split in two, one half two or three kilometers above the other.

Mercury is a world created by the sun, and left only partly finished.

NASA

Mercury's south pole is located inside the large crater at the top, on the terminator (above).

Ron Miller

A cliff, or scarp, more than 300 kilometers long stretches diagonally across Mercury (above), splitting more than one crater in two.

Nothing got in the way of the great shifting blocks of Mercury's crust. The scarps split mountains and craters alike. Here (above), Discovery Rupes bisects a crater (demonstrating neatly that the scarps are younger than this crater), so that one half of it is two kilometers above the other half, resembling a broken dinner plate.

The largest crater in this ray-striped view (above) is 100 kilometers across.

One of the features that distinguishes Mercury from our moon are the enormous scarps, or cliffs, that stretch for hundreds of kilometers across its coppery surface. These scarps—or in the Latinized international terminology, rupes—are four kilometers high. They are the result of massive blocks of Mercury's crust being thrust upward along fault lines, probably while the early crust was still cooling and shrinking. This particular cliff (above) is some 500 kilometers (310 miles) long and is called Discovery Rupes.

Mariner photographed one-half of Mercury's Caloris Basin (above), the series of concentric two-kilometer-high mountains and ridges on the left side. It is the 1,300-kilometer result of the collision of Mercury with a small asteroid. It has relatives on both our moon and Callisto.

105

MERCURY

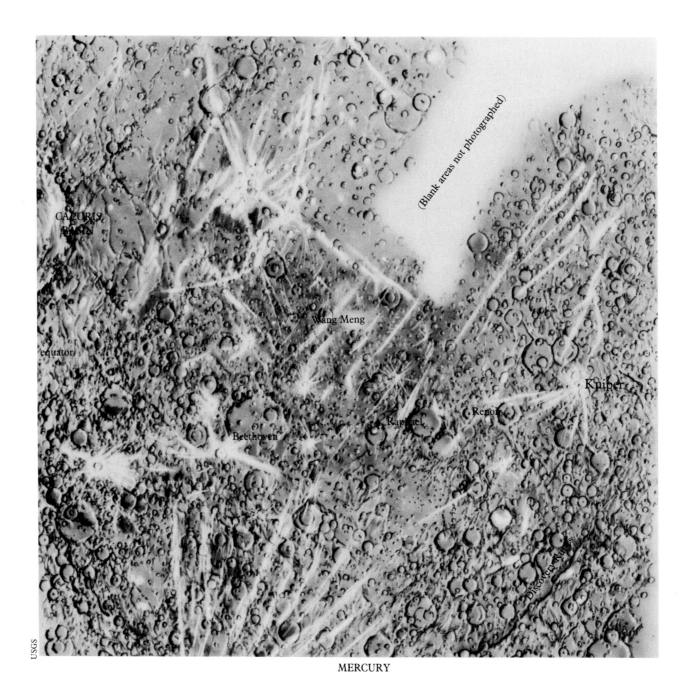

USGS

TRITON A WORLD OUT OF PLACE

Distance from Neptune: 355,000 km.
Revolution: 5 days 21 hours
Diameter: 4,400 km.
Composition: ices (?)

Among the planetary-scale moons of the solar system, Triton lies just beyond our current frontier in several senses. The inner of the two moons of Neptune, it is too far away for us to know much about it, and it has not yet been visited by our space vehicles. However, we do know that it is quite large—a recent tabulation places its diameter at 3,200 to 4,800 kilometers or about 2,000 to 3,000 miles. This would make it the fourth largest satellite in the solar system, after Ganymede, Titan, and Callisto; and would also make it about the size of the planet Mercury and perhaps 30 percent bigger than its neighboring "planet" Pluto (whose "planetary" status is somewhat in doubt).

Spectroscopic observations have failed to determine the surface material on Triton. Most likely, judging from its −223°C (−360°F) temperature, the material is dominated by water or methane ice, perhaps combined with some rocky soil.

Triton attracted attention in 1979 when astronomers announced that spectroscopic observations had revealed that it showed faint traces of a thin methane atmosphere, with a pressure only one ten-thousandth that of Earth's atmosphere. This would not be noticeably different from the vacuum of space to astronauts on Triton's cold surface. A certain amount of methane gas would be expected to sublime off the methane ice presumed to be on the surface, and the reported amount of gas is consistent with this assumption.

The methane atmosphere is so thin that its importance lies more in confirming the presence of methane ice on the surface (as a source for the gas) than in its presence as an atmosphere. It is also remotely possible that Triton is big enough to have internal geologic processes that could occasionally emit methane as a volcanic gas. Most such gas emissions would leak off into space because of Triton's weak gravity, so that the amount of atmosphere could vary depending on how long it had been since the last eruption.

Triton's orbit presents a more fundamental enigma. Triton is the only large satellite with retrograde orbital motion (east to west, or clockwise as seen from the north), opposite to that of the planets and most other satellites. Also, Triton has an orbital inclination of 20 degrees;

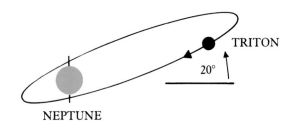

TRITON

20°

NEPTUNE

TRITON

no other large satellite except the moon has an inclination higher than 1 degree. The effect of tidal forces caused by Neptune on such a satellite is to make the satellite spiral slowly inward toward its planet. Because of these forces, Triton may spiral inward and crash into Neptune within the next 100 million years. This is a startling fact; cosmically, 100 million years is an extremely brief period of time, considering that the solar system has been stable for 4.5 billion years.

All of this suggests that something strange may have happened to Triton during its history. For example, some astronomers have proposed that Pluto did not form as a planet, but rather as another satellite of Neptune, and that the three interacted closely in a way that changed their orbits, threw Pluto out of the system, and caused Triton to embark upon its unusual orbital path. This is possible but not very likely, especially since Pluto has its own satellite, which would have to have traveled with it from the Neptune system. Another possibility is that a giant comet from the great swarm of comet nuclei in the outer solar system

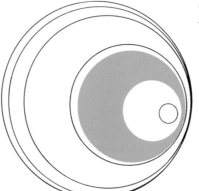

Triton (solid color) compared in size to (from largest circle to smallest): Ganymede, Titan, Mercury, The Moon, Charon and Mimas.

passed through Neptune's satellite system and disrupted it, changing Triton's orbit from a more normal one to its strange, suicidal course. Still another alternative is that Jupiter or Saturn perturbed a large asteroidlike body into a path through Neptune's satellite system, causing the disruption.

We are a long way from understanding Triton, but its cold, icy surface seems likely to harbor some interesting secrets about the history of planets and satellites.

Ice-rimmed fumaroles silently smoke beneath the sea-green globe of Neptune (right). Some researchers suggest that there are thin amounts of methane gas on the surface of Triton. This gas may sublimate off methane ice on Triton's surface. However, Triton is large enough, and has had such a disrupted history, that its interior may occasionally vent small amounts of "volcanic" gases, perhaps derived from melted methane or water ices in its interior. If Triton once underwent a close approach to another large body that changed its orbit, as some researchers suggest, it may have been heated and intensely fractured, generating and emitting internal gases. The planet Neptune, around which Triton orbits, covers an angle in the sky of about 8 degrees (sixteen times that of a full moon).

Ron Miller

THE MOON EARTH'S COMPANION

Distance from Earth: 381,575 km.
Revolution: 27 days 7 hours
Diameter: 3,476 km.
Composition: rock, silicate

The moon has come to be one of the best-known worlds. On July 20, 1969, *Apollo 11* astronauts became the first human beings to set foot on the moon. The samples they and several Russian automated landers brought back have revealed a great deal about the evolution of the moon.

Lunar rocks are generally older than those found on Earth. For this reason, lunar exploration has been geologically valuable, helping to fill in the gaps in the first billion years of the Earth-moon system's history.

What the moon rocks tell us is that the moon (and probably Earth) began its history 4.5 billion years ago with a surface layer that was molten. This molten layer, or *magma ocean*, as it is called, was a few hundred kilometers deep. As it cooled, various minerals solidified and a crust of lower-density minerals aggregated on the surface, accounting for the low-density rock types known as *anorthosites*, which form much of the ancient lunar highlands. By 4.4 billion years ago, much of the magma ocean had solidified, although deep pockets of molten material existed under the surface.

A small lunar crater (below) photographed by Apollo 12 astronauts.

Counts of lunar impact craters and determination of rock ages tell us that the rate of meteorite impact was extremely high between 4.4 and 4 billion years ago. This was probably true on all planets throughout the solar system. Meteorite debris was left over from the formation of planets 4.5 billion years ago; scattered through interplanetary space, this debris often collided with them. But by 4 billion years

We see a face of the moon never visible from Earth as we hover 8,000 kilometers (5,000 miles) above the lunar far side (right). We miss the familiar maria—the dark patches that make up the "man in the moon." Instead, there are only endless, wall-to-wall craters. The one with the prominent dark floor is Tsiolkovskij, named for the great Russian rocket pioneer. In this view we are positioned beyond the orbit of the moon, with Earth looking four times larger than the moon does from Earth.

MOON

W.K. Hartmann

ago, much of the debris had been "collected" by the planets. The impact rate was so great before then that only chips and pieces of lunar rock survive from that time.

Around 3.9 billion years ago, heat generated by radioactivity had accumulated inside the moon, melting a layer several hundred kilometers beneath the surface. The largest meteorite

We see a relatively fresh lava flow on the moon (left). Dates of numerous rock samples from the lava plains of the moon show that eruptions of lunar lavas were strongly concentrated between 3.2 and 3.8 billion years ago. The outer layers of the moon probably cooled to below molten rock temperatures around 3 billion years ago. Lava flows as old as that have been "sandblasted" by the impact of small meteorites until their surface textures are obliterated in a layer of dust and rocky rubble.

It seems unlikely, however, that all pockets of underground lava disappeared in all regions of the moon 3 billion years ago. Some researchers have estimated that certain fresher-looking flows are only about 2 billion years old. Other researchers suspect that vestiges of volcanic activity, such as gas emission, occur even today. Perhaps we will eventually locate some lunar region where a small, localized flow only 1 billion or .5 billion years old has been relatively well-preserved.

No large volcanoes occur on the moon; instead, lava flows appear to originate from fissures and low ridges. The lunar lavas were probably too hot and too fluid to pile up into steep-sided mountains.

impacts, which formed huge craters of a type known as multi-ring basins, caused fractures that penetrated into this magma layer. Lava ascended to the surface and erupted, covering the floors of the huge basins and some of the larger craters. These lava flows are darker-colored than the ancient highlands, and form the dark plains that we see from Earth—the "man in the moon." Galileo and other early astronomers erroneously thought that these flat areas were oceans, and gave them the Latin name for seas, *mare* (plural *maria*.) Thus, these features bear colorful names, such as Mare Tranquillitatis (Sea of Tranquillity) and Mare Imbrium (Sea of Showers).

Most of the prominent maria, or lava plains, formed between 3.8 and 3.2 billion years ago.

Shadows cast by the setting sun point toward a nearly full Earth, sitting on the lunar horizon like a marbled bowling ball (right). The elongated, jagged shadows of the mountains demonstrate why for so many years people—astronomers and laymen alike —were misled into thinking that the mountains themselves were craggy and jagged. This impression was created by low lighting angles. The sun rises and sets, although a lunar day seems intolerably long—fourteen Earth days from dawn to dusk—but Earth never moves from its place in the sky, since the moon always keeps one face toward us.

MOON

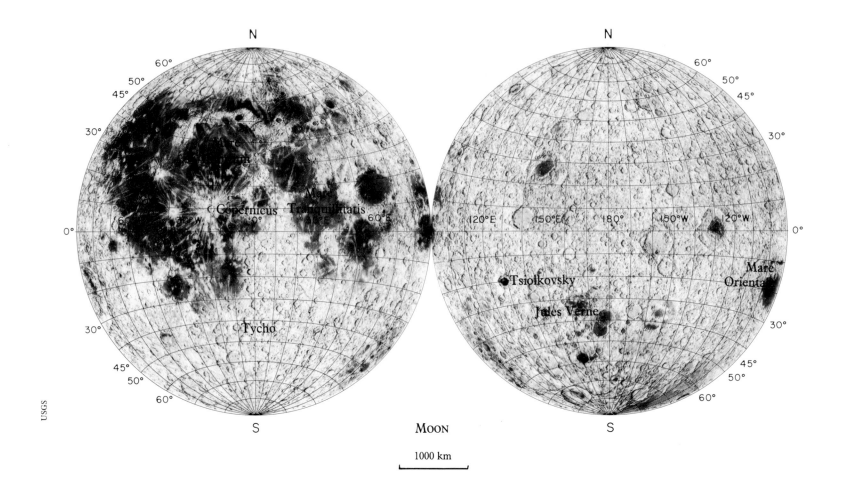

Copernicus Mare Tranquilitatis

Tycho

Tsiolkovsky

Jules Verne

Mare Orienta

MOON

1000 km

USGS

Their rocks are basaltic lavas, similar to the volcanic lavas found in Hawaii, California, Arizona, and elsewhere.

As the Earth evolved its oxygen atmosphere, plants, and animals (3.2 billion years ago), the airless moon continued to be belted by a rain of meteorites. Numerous small ones sandblasted the surfaces of the rugged lava flows into smooth, undulating layers of fine dust and rock chips about ten meters (thirty feet) deep. This was the surface our astronauts trod, kicking up dust that dirtied their space suits and sprayed from their rovers' tires in graceful arcs. This dusty layer is called the lunar *regolith*.

Larger meteorites fell at a rate only one one-thousandth as often as they did a billion years before. Their occasional impacts excavated craters through the regolith and ejected rock fragments from the lava layers below, accounting for scattered rocks in the maria plains and the uplands.

Very large impacts blasted out rare, large young craters such as Tycho, whose rays of bright ejected debris stretch outward across the moon. Interiors of such craters have not been sandblasted enough to be smooth, and so are marked by chaotic rubble that must form landscapes far rougher than any we have seen.

Until 1965, heavily cratered landscapes like those of the moon were unknown elsewhere in the solar system because our ships had not yet reached other planets. Today, we know that virtually all the twenty or so planetary surfaces we have seen at close range are cratered, and that the moon displays the effects of processes that have acted on many worlds throughout the solar system. Orbiting spacecraft have shown us that even Earth is more heavily cratered than had even been suspected; erosion and plant life have made even enormous craters invisible from the ground.

We have landed on the moon in only a few places, and those have been relatively flat. We have not yet seen the rugged areas, the fresh craters, the nighttime surface illuminated by Earthlight, or the lunar landscape bathed in the fiery red light transmitted through Earth's atmosphere during an eclipse, when Earth passes between the moon and sun. Many vistas remain for us to admire on our neighboring world.

NASA

Zooming in on Tycho (left above, middle, below), *the final photograph covers an area about 5 by 6.5 kilometers.*

Forbidding badlands convolute the floor of the crater Tycho (overleaf), *bound by the distant, towering cliffs of the crater walls. This ruggedness is an indication of Tycho's relative youth—it has yet to be sandblasted by micrometeorites to match the rolling smoothness of much of the rest of the moon. The three photographs* (left) *zoom us into Tycho, to show us the area in which we're now standing. The crater, the result of a meteorite collision, is the size of Yellowstone National Park. Tycho's central peak is a mountain about 1800 meters (6000 feet) high. The crater walls, which rise in a series of stepped terraces, loom 3600 meters (12,000 feet).*

115

Tycho

118

Rubble at the foot of Mt. Hadley (right), photographed by Apollo 15 astronauts.

In the rough country around the Orientale impact basin, Earth always hangs low over the eastern horizon. Once the sun sets (left), eerie blue Earthlight is the only source of illumination. Here it catches the top of one of the huge cliff-rings that surround the Orientale Basin. Because this basin is hundreds of kilometers across, we do not have the sense of standing in a crater; rather, the fault-ring appears more like a long, scalloped mountain range. The sun is just below the western horizon, which we face. Since there is no air, there are no glorious sunset colors; but the moon offers a different kind of sunset display. The horizon glows with the outer parts of the sun's corona and a faint band of zodiacal light—the dust of the inner solar system lit by the sun.

Because the moon turns in twenty-nine and a half days (relative to the sun) instead of twenty-four hours, the sunset is very slow. The sun actually hangs just below the horizon for an entire Earth-day of twilight before the full, fourteen-day lunar night sets in. But the cliffs are still lit by Earth. They will fade as Earth goes from its full phase (which it is in now, since it is opposite the sun) to a much fainter crescent illumination just before sunrise.

A close-up view of a crater wall (above). The distance from the crater's central peak, at the bottom, to the rim is about thirty kilometers. Material has slumped from the wall onto the crater floor, which is covered with domes and low ridges.

Looking northwest up Hadley Rille (left), a relatively shallow, meandering valley like many others on the moon.

119

Ron Miller

NASA

Apollo 15 astronauts photographed this boulder (below), thrown out of a nearby crater when it was formed, as were most of the other rocks around it.

As far as the human race is concerned, this is probably the most historic site on the moon (left). It is here that mankind, through its emissaries, first reached out and touched another world. On July 20, 1969, Neil Armstrong and Edwin Aldrin, Jr., made the first human footprints on an unearthly landscape, while Michael Collins patiently orbited overhead in the command module. When they departed one and a half days later, with scores of photographs and rock and soil samples, they left this monument on the lava plains of Mare Tranquillitatis. Barring accident, it will last for eons. Even something as fragile as the astronauts' footprints will remain unchanged for millenia, only filling in imperceptibly via the unending dusting of micrometeorites. The spiderlike structure behind which the sun is setting is the lower stage of the lunar module Eagle.

A gibbous Earth hangs brilliantly above the rugged mountains of the moon (right). There are several ranges of peaks, all named after terrestrial mountains: lunar Apennines, Alps, Cordilleras, and Pyrenees. The relatively gentle slopes and rolling profiles are misleading: many of these lunar mountains tower far higher than their terrestrial namesakes. Some rise as high as 4,000 or 6,000 meters (13,200 and 19,800 feet) above the lunar plains. Mont Blanc, highest peak of Earth's Alps, is only 4811 meters high. Millions of years of being scoured by micrometeorites—and a few larger ones—have reduced most of these massive ranges from their original Rocky Mountain cragginess to a lofty but Appalachian gentleness.

Ron Miller

MOON

This large, crater (below) is on the far side of the moon, and is about eighty kilometers wide.

NASA

Rugged boulders at Taurus-Littrow (below).

NASA

The surface of the moon has turned to copper as the blood-red light from an eclipsing Earth washes over the landscape (right). The backlit atmosphere of Earth, as it passes in front of the sun, is illuminated in an orange-red ring. Sunlight, passing through the atmosphere at such a low angle, turns red just as it does at sunrise and sunset. The reddened sunlight refracted through Earth's atmosphere is what is lighting everything around us.

Pale streamers of the sun's corona fan out from behind Earth while the even paler zodiacal light extends beyond them in a faint band. The sun is about to pass out of eclipse; in fact, the first sunlight is just striking the horizon beyond Earth's shadow.

Ron Miller

The Mare Orientale basin as photographed by Lunar Orbiter (left). Its outer ring, the Cordillera Mountains, are over 965 kilometers in diameter. Mare Orientale is probably the result of the impact of a small asteroid.

Even though Earth is fixed in the lunar heavens, the sun moves, and as it does, Earth goes through all the familiar phases: crescent, half, full, half, crescent and new. Earth is in crescent phase here (right), its night side faintly lit with a ghostly light. From Earth, the moon always appears in the phase opposite that which Earth is in seen from the moon. Since Earth is crescent here, the moon is gibbous from Earth. Sunlight reflected from the moon is lighting Earth's night side. The soft glow is moonlight, shining on Earth's clouds and seas.

125

PLUTO THE DOUBLE PLANET

Average distance from the sun:
5,900,000,000 km.
Length of year: 248 years
Length of day: 6 days 9 hours
Diameter: 3,000 km. (?)
Surface gravity (Earth = 1): .05 (?)
Composition: ices (?)

Pluto was discovered in 1930 by Clyde Tombaugh, an astronomer at Lowell Observatory in Flagstaff, Arizona. It was immediately hailed as the ninth planet, the only planet to have been discovered in the twentieth century—a designation it has maintained in astronomy textbooks until today. However, some interesting properties of Pluto have recently come to light, calling into question whether it shares the same mechanisms of origin and evolution as the other planets.

For one thing, Pluto is smaller than any other planet, about the size of our moon. Present estimates put it in the range of 3,000 to 4,000 kilometers (1,860 to 2,480 miles) in diameter. A likely estimate of 3100 km, based on recent measures, would make Pluto only a bit over three times the size of the largest asteroid, Ceres. (Ceres was at first classified as a planet, but later downgraded because it was so small, and because it is accompanied by numerous smaller objects in neighboring orbits—the other asteroids.) Pluto has been known for years to be in a region populated by

comet nuclei, but most of them were thought to be much smaller than Pluto. However, in 1977 the unusual object Chiron—perhaps 10 percent the size of Pluto —was discovered in the region between Uranus and Saturn.

Both Pluto and Chiron have orbits that cross over the orbits of other planets. Pluto sometimes travels inside the orbit of Neptune: in fact, it will be closer to the sun than Neptune is during the 1980s. In this sense, Pluto and Chiron are more like asteroids than planets, but they are far outside the part of the solar system inhabited by conventional asteroids. These facts raise the possibility that Pluto may come to be regarded as only the largest of a swarm of

A heatless spark, the sun glimmers off crystals of methane, ammonia, and other gases—frozen hard as steel (above).

bodies in the outer solar system —some as yet undiscovered— ranging downward in size to the objects normally catalogued as comets. To some extent, this is just a question of semantics; but the way we look at our solar system may enable us to ask the right questions—and therefore learn the most—about our environment.

In 1977, an astronomer at the U.S. Naval Observatory discovered that Pluto has a satellite, which they named Charon (not to be confused with the unrelated object Chiron), a

Ron Miller

As pale as a moonlit skull, gray Pluto hangs in the sky of frigid Charon (above), its giant moon and only companion as it ponderously rolls through the comet-haunted darkness at the edge of the solar system. So close do the two huddle that Pluto appears seventeen times the size of a terrestrial full moon.

composed of ices perhaps less dense than even water ice. One likely component is methane, which would freeze only in the outermost parts of the solar system. The existence of frozen methane on Pluto might suggest that Pluto is native to the outer solar system.

Direct proof that methane exists on Pluto's surface comes from spectroscopic observations. In the latter half of the 1970s, various astronomers reported spectral absorptions that they attributed to methane. A study in 1980 proposed the existence of methane ice on the surface of Pluto, and calculated that some of this ice ought to sublime into gaseous methane, providing a thin atmosphere of methane gas when Pluto is closest to the sun. A subsequent 1980 study detected a thin Plutonian atmosphere of gaseous methane phere of gaseous methane with only one ten-thousandth the surface pressure of Earth's atmosphere. This study also concluded that some previously reported features were due to methane gas, not methane ice on the surface.

Whatever the resolution of these varying observations, it appears that Pluto must have both an imperceptibly thin methane atmosphere *and* methane ice on its surface that is the source of the gas.

At forty times our distance from the sun, Pluto is so remote

mythical underworld associate of Pluto. The satellite is about half Pluto's size—a much larger size ratio than found between any other planet-satellite pair.

The fact that Pluto has its own satellite also might tend to elevate it toward planetary status, except for recent evidence that indicates that some asteroids have satellites as well.

Recent observations of Pluto's mass and size indicate that its mean density is very low. This means that Pluto is mostly

that the sun's disk seen from its surface covers less than a minute of arc—so small that it looks like an intensely bright star. Its light is only one fifteen-hundredth of the brightness of sunlight on Earth, but is still 250 times brighter than the light of a full moon on Earth. The appearance of the sun might be likened to that of a very brilliant streetlamp a little way down the block. On Pluto, it is easy to believe that the sun is only just one more of the many stars in the night sky.

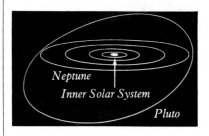

Neptune
Inner Solar System
Pluto

The sun is little more than a brilliant star when seen from the ice-bound surface of Pluto (overleaf). Orbiting in the farthest frontier of our solar system, accompanied only by its giant moon, Charon, Pluto knows only the cold of interstellar space. We are standing at the edge of a lake of frozen methane that dimly reflects distant cliffs, glinting like glaciers in moonlight. Forty times as far from the sun as the Earth, Pluto receives far less warmth and light than Earth. Charon is a gray orb, lit wanly by the heatless sun.

127

Pluto

129

EUROPA AN ICY CUE BALL

Distance from Jupiter: 671,000 km.
Revolution: 3.6 days
Diameter: 3,130 km.
Composition: ice

With a diameter of 3,130 kilometers, or 1,941 miles (slightly smaller than our moon), Europa is the fourth largest of Jupiter's satellites. Even before the two *Voyagers* sailed by, spectra obtained by Earthbound astronomers showed that Europa's surface is composed mostly of frozen water. These spectra and other observations indicated that Europa's surface is brighter and richer in ice than the other satellites, which (in the cases of Ganymede and Callisto) are made of a mixture of ice and rocky soil.

Voyager revealed that Europa is the most nearly featureless world known in the solar system. It looks like a mottled, cream-colored cue ball, with only faint markings and virtually no large impact craters. The markings are primarily a set of tan streaks with so little relief that they look as if they had been drawn on Europa with a felt-tipped pen. The streaks are only about 10 percent darker than the surface—enough contrast to be clearly visible but not bold; many *Voyager* TV photographs that show the markings as prominent dark streaks have been processed for contrast. Europa's streaks actually resemble the fictitious "canals" that Percival Lowell erroneously drew on Mars. A *Voyager* scientist, seeing the first images of Europa, asked plaintively, "Where is Percival Lowell now that we need him?"

What are the streaks of Europa? Many seem to be broad, shallow valleys, about five to seventy kilometers (three to forty-three miles) across and stretching as far as three thousand kilometers (1,860 miles) in straight or curved paths across Europa's plains. Some of the wider ones have brighter central strips, perhaps flat valley floors. To explain this, *Voyager* scientists studied photos of Europa's terminator—the zone dividing day from night, where even gentle relief can cast long shadows. Even here, most streaks revealed no visible shadows, suggesting the valleys (if they are valleys) are less than a few hundred meters deep. Among the darker streaks along the terminator, one can see bright ridges about ten kilometers (six miles) wide, which rise no more than a few hundred kilometers above the surface. These low ridges often have scalloped patterns, with each scalloped curve running a hundred kilometers or so along the plains.

The explanation of these features involves both the composition and the history of Europa. Although Europa's *surface* is essentially ice, the mean density of Europa's *interior* is much greater than that of ice—it is about that of silicate rocks, such as volcanic lavas on Earth or the moon. This implies that Europa is mostly a rocky world, with only a layer of ice (perhaps as much as a hundred kilometers [sixty-two miles] thick) covering the surface. The absence of large craters (only three larger than twenty kilometers [twelve miles] across have been mapped) implies that Europa's surface does not date back to the era of intense cratering, 4 to 4.5 billion years ago, when the planets were formed. The icy surface layer was probably created more recently. Perhaps the ice layer flowed like a glacier, filling in the older, deeper craters.

Like Ganymede and Callisto, Europa probably formed from a mixture of icy and rocky dust in the cold, primordial cloud of debris orbiting around newly formed Jupiter. But Europa received more heat than the outer two satellites. This radiant heat may have come from massive Jupiter, or from tidal flexing, the mechanism that melted the inner satellite, Io, and produced its volcanoes. Wherever it came from, it was sufficient to melt portions of Europa's interior, allowing watery "lava" to erupt and coat the surface. As the surface cooled, it probably looked much like our Arctic Ocean, with thick pack ice floating on a watery sea. Currents

NASA

Sea ice breaking up off the coast of Antarctica (below) may be analogous to the fracturing of Europa's early frozen crust.

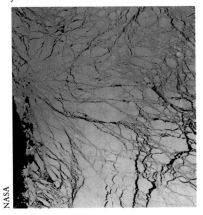

NASA

Veined like a cracked ivory ball, Europa (right) presents the flattest surface known in the solar system. The lines are grooves, or fractures, in the icy crust.

NASA

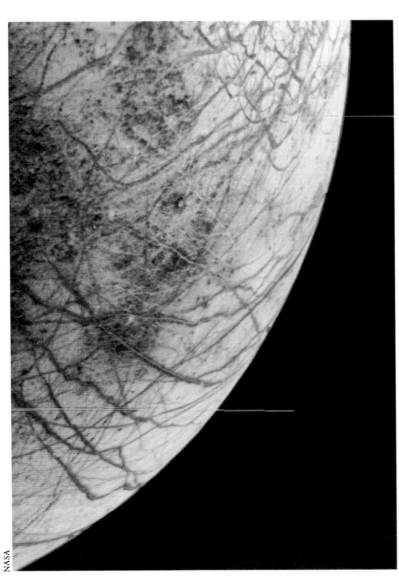

in the sea stressed the ice, causing fractures. Indeed, close-up photos of Europa look strikingly reminiscent of aerial photos of cracked ice sheets in the Arctic. Over the years, eruptions of fresh water through the cracks froze, converting the fracture system into a pattern of shallow valleys and bright ridges. The processes were similar to those that made the fractured zones of neighboring Ganymede. However, there was less heat than on neighboring Io, so volcanic eruptions did not drive off all the water, build volcanic mountains, or create sulfurous lava flows.

Monolithic crystals of water ice have burst through the floor of a broad fracture in the ice ocean that covers Europa (overleaf). The fracture itself is a broad, shallow valley, its low walls at the far left and far right. Fractures make Europa look like a shattered crystal globe.

Ice at moderate temperatures is not very rigid. As Earthly glaciers demonstrate, ice will flow under its own weight. Any large fractures Europa may once have had have therefore long ago disappeared. In fact, Europa may be the smoothest object in the solar system: if it were the size of a billiard ball, it would be smoother than a billiard ball.

Europa: Ice Fracture

EUROPA

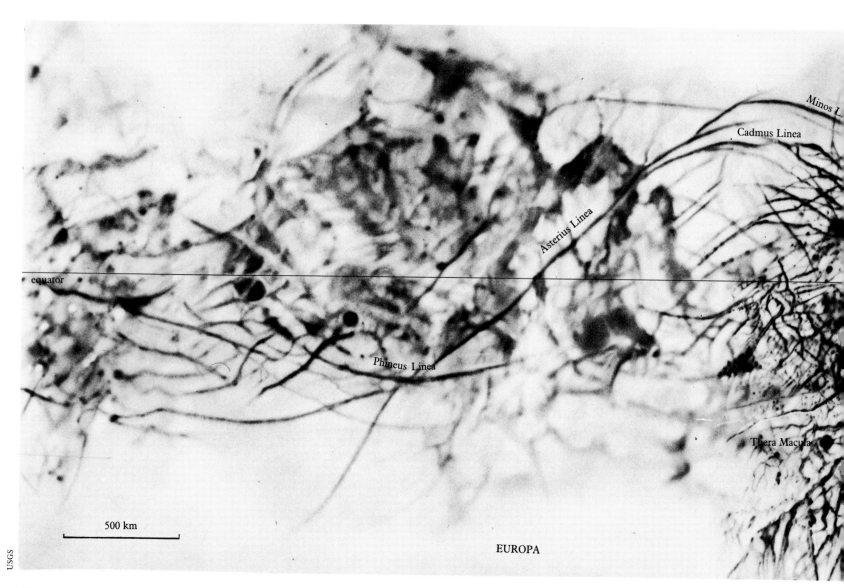

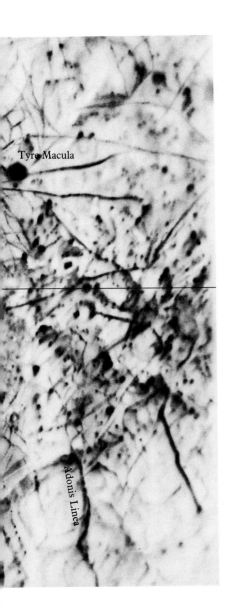

Tyre Macula

Adonis Linea

TITANIA

Titania is the fourth of five Uranian satellites, in order outward from the planet. Its diameter is uncertain, but it's probably about 1,000 kilometers (620 miles) making it the largest and brightest of Uranus's moons.

There are spectroscopic signs of frozen water on the surface of Titania. This is not surprising. Jupiter's satellites are somewhat icy, and Saturn's even more so; in the cold depths of the outer solar system, we can expect a still higher proportion of ices to have condensed. The Uranian moons, however, differ from Saturn's in appearance. Perhaps some other component, possibly methane, is mixed into Titania's ice. Or maybe Titania's ice fields have some different physical form—powdery snow, frost crystals, or some low-temperature form of ice with unusual crystals—than those on Saturn's moons.

The gravity on Titania is too slight to hold gas atoms very long, so it and its neighboring moons lack any detectable atmosphere.

Looking like a sea-green Christmas tree ornament, Uranus appears above the ice-covered surface of Titania (overleaf). The night side of the crescent planet is lit dimly by light reflected from its coterie of five moons. The three moons that are inside Titania's orbit can be seen in the same plane as Uranus's faint rings. Two are between us and the ring; the large one is Umbriel, the other is Miranda. The moon just beyond the edge of the rings is Ariel. Uranus's fifth moon, Oberon, is outside Titania's orbit, behind us.

Uranus's tipped axis provides us with some unusual phases. Here the planet's north pole is in sunlight, the south pole in darkness. The terminator, contrary to our experience on other planets, runs from east to west rather than from north to south.

Ron Miller

Uranus from Titania

RHEA

Rhea is a little less than half the size of our own moon—only 1,530 kilometers (950 miles) in diameter. It is in the middle of Saturn's satellite system, one in from Titan. Spectroscopic observations show that it has no atmosphere at all and that its surface is a bright material reflecting 60 percent of the sunlight falling on it—mostly frozen water and a small amount of rocky soil. This surface, eroded by meteorite sandblasting, looks something like a rolling plain of finely crushed ice.

Close-up photos taken by *Voyager 1* reveal a wilderness of craters; Rhea is one of the most crater-crowded of any world in the solar system. It looks as though the bombardment has gone on at random since the formation of the planets, 4.5 billion years ago. Few craters have been destroyed or eroded by new impacts. If this is true, it suggests that Rhea has never had much internally-generated geologic activity or tidal heating, either of which would have altered Rhea's pockmarked surface.

The densities of Rhea and its immediate neighbors are relatively low, implying more ice and less rock than in Jupiter's large satellites. Because of the lower temperatures at this distance from the sun, worlds that formed here may have contained more ice than those that formed near Jupiter. Rhea gives us a good opportunity to learn about the evolution of a world composed mainly of ice.

RHEA

500 km

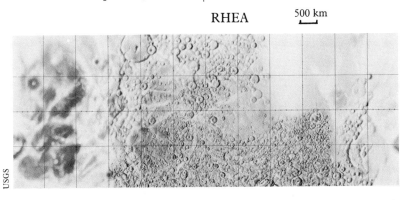

USGS

The heavily cratered surface of Rhea (right). Some of the craters are as large as seventy-five kilometers. Many have sharp rims and appear to be fresh, while others are shallow and degraded and are probably very old.

One of the most heavily cratered regions on Rhea is visible in this photograph (below). The smallest features seen are about 2.5 kilometers across. Sun glints off fresh ice exposed in the walls of some of the craters.

NASA

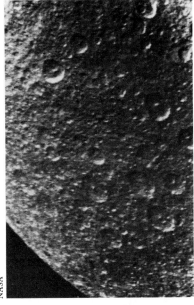

NASA

Saturn, seen from the surface of Rhea (overleaf), is a huge globe nearly twenty-six times the size of a full moon back on Earth. Wreathed in its brilliant rings, Saturn presents an ever-changing display unparalleled anywhere else in the solar system. Like most of its sister moons, some of which can be seen in the plane of the rings, Rhea's surface is made of frozen water. Walking across its crunchy, undulating landscape would be like crossing a glacier on Earth—though this glacier covers the entire world. There are sixteen moons circling Saturn. If you lived there, you could choose between sixteen different lengths of month, ranging from less than an Earthly day to more than a year.

Rhea

IAPETUS

Iapetus, an outer satellite of Saturn, has been a puzzle since its discovery in the 1600s. Early telescopic observers were astonished to find that they could see the tiny moon when it was on one side of Saturn, but not when its orbit carried it around to the other side of the planet. Later observations revealed why. Like our own moon—and indeed like all satellites in the solar system seem to—Iapetus keeps one face toward its parent planet. Iapetus's trailing hemisphere, which faces Earth when it is on one side of Saturn, is as bright as ice and very visible. But the leading hemisphere is only one-fifth as bright—as dark as bare rock. It is simply too dark and faint to have been seen in early telescopes.

Visiting Iapetus is visiting a world divided. One side is surfaced with frozen water, the other with blackish, rocky material. The boundary between the two is fairly sharp. Only the dark rim of a giant ice-filled crater mars the brilliant ice-covered hemisphere. From either side, Saturn provides a magnificent sight: a pale yellow ball, four times larger than a full moon back on Earth, surrounded by dazzling white rings. With the exception of even more distant Phoebe, Iapetus is the only one of Saturn's moons that allows the visitor to see the rings as a wide band, instead of as a thin white line. This is because Iapetus's orbit is tipped almost 15 degrees, so that it passes well above and below the plane of the rings. Traveling across Iapetus, we see Saturn above a landscape of rolling ice. Then, suddenly, we cross the border into the other hemisphere, and Saturn seems to glow even brighter in contrast with Iapetus's desolate rocks.

A possible explanation for this two-faced moon involves dust blown off of Phoebe, the next and farthest moon out. Phoebe is dark gray and, alone among Saturn's moons, moves in a retrograde orbit. This means that any debris knocked off it by meteorites circles Saturn in the same clockwise direction. Dust from Phoebe tends to spiral inward and eventually encounter Iapetus. But the encounter is at high speeds because of their opposing directions. Thus, Iapetus gets hit by high-speed dark dust coming at its leading side. This probably not only coats that hemisphere with a layer of dust, but vaporizes the ice in the original soil.

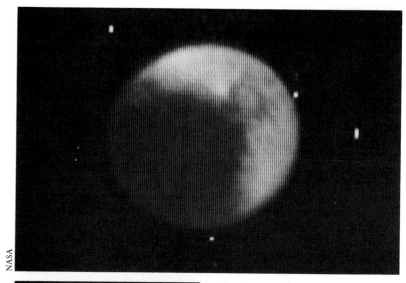

NASA

Ron Miller

Saturn seen from Phoebe (above).

The large circular feature (above), *about 200 kilometers wide, is probably a crater with a dark central mountain. This photograph was taken at a distance of 3.2 million kilometers by Voyager 1. Iapetus's leading hemisphere is to the left and is four or five times darker than its other hemisphere, an asymmetry first noted from Earth.*

Ron Miller

Saturn looks very much like a diamond ring, hovering above the icy landscape of Iapetus (above). Iapetus and Phoebe are the only two of Saturn's moons whose orbits are inclined to the plane of the rings. During the course of its eighty-day trip around Saturn, Iapetus first swings "above" and then "below" the planet. Here we see Saturn at its maximum tilt as we reach the "highest" point in Iapetus's orbit. Seen from almost ten times the Earth-moon distance, the giant planet still takes up as much sky as a full Earth seen from the moon. We're on Iapetus's ice-covered hemisphere; the other side of the moon has a surface of dark soil.

142

ARIEL & MIRANDA

Named for the mischievous imp in Shakespeare's The Tempest, *tiny (910-kilometers or 488-mile) Ariel orbits only 191,800 kilometers (116,000 miles) above the cotton candy globe of Uranus (left), whose ring seen edge-on as a thin gray line makes it look like a blue-green derby. Uranus's northern hemisphere is in the middle of its twenty-one-year-long summer. The north pole is pointed directly at the distant sun; the south pole is in the midst of a long, dark winter. Skewered by the pale ring is innermost moon Miranda.*

Ariel is just small enough so that it is not one of the worlds that partially melts inside; instead, it will remain frozen and solid forever. Therefore, its surface has never been seriously disrupted by eruptions or faulting, and is covered by impact craters that have remained unchanged down through the eons.

Ron Miller

Uranus (above) swells like a green balloon above the frigid surface of Miranda. The tiny moon is most likely a solid sphere of ice, 550 kilometers (341 miles) in diameter. The shrunken sun, looking twenty times smaller than it does when seen from Earth, provides scarcely one four-hundredth of the heat and light we enjoy.

Uranus is in mid-spring, when its phases most resemble those of the rest of the planets. Its rings and equator are nearly at right angles to the plane of the solar system . Its crescent runs from north pole to south pole. For the next ten and a half years, both northern and southern hemispheres will enjoy both light and darkness during the course of the thirteen to twenty-four-hour day (its exact length is uncertain).

CHARON

Charon, Pluto's only known satellite, was discovered in 1977 by U.S. astronomer James Christy. He noticed that photographic images of Pluto weren't round—the planet seemed to have a "bump." Further photographs showed that the bump kept changing position: a satellite was orbiting Pluto.

Earlier photographs of Pluto also showed the odd elongation. It is possible that someone, looking through the plates, discarded the ones showing the misshapen Pluto, concluding that the strange bulge was a blur due to a movement of the telescope.

Charon's existence makes Pluto a more interesting place. First of all, it gives Pluto the biggest satellite, relative to its own size, of any planet. Charon is 40 to 50 percent as big as Pluto. (The moon is only 27 percent the size of Earth, and Triton in only 9 percent of Neptune's size.) Secondly, Charon has the largest angular size of any moon seen from its planet. Seen from Pluto's

The Pluto-Charon distance to scale.

Pluto-Charon Comparison

surface, Charon is about 4 degrees wide, eight times the size of a full moon on Earth.

Charon is about 1,300 kilometers (800 miles) in diameter, quite a respectable size. Very little is known about its physical nature because it is so hard to visually separate from Pluto. Its surface is probably similar to Pluto's—mostly methane ice. Charon and Pluto are a unique double planet, orbiting each other every 6.4 days in a stately pas de deux.

We are hovering behind the unlit nightside of Charon (right)—surely one of the most cheerless places in the solar system. The sun, just behind our field of view at the right, is a virtually heatless point of dazzling light, 1,600 times smaller than when seen from Earth.

Charon and Pluto swing through space like a pair of ponderous waltzers. Locked face-to-face, neither ever appears to move from its place in each other's sky. Only the stars in the background seem to move, coming full cycle every 6.26 Earth days. Charon, in proportion to Pluto, is an enormous moon, about half the size of Pluto; about 19,000 kilometers (11,800 miles) separate the two. So large are they in proportion to each other, and so close, that both orbit around a third point—called a barycenter—between the two. Earth and the moon do the same, but their barycenter is still within the body of Earth, though quite a bit away from the center.

Here, in the most distant frontier of the solar system, it is as still as death. Water, even oxygen and carbon dioxide, are as hard and brittle as glass. The shores surrounding pools of liquid hydrogen are undisturbed by their weightless ripples. Charon has no atmosphere and its sky is crystal clear, but the brightness of even the highest noon is less than an Earthly twilight, and the stars never cease to shine.

TETHYS & "JANUS"

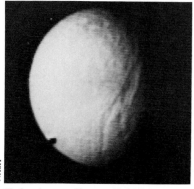

NASA

Tethys is a small moon (1,200 kilometers in diameter) orbiting between Dione and Enceladus, about 30,000 kilometers from Saturn. This is a Voyager *photograph of the side of Tethys that faces Saturn. It is scarred by an enormous trench some 750 kilometers long and 60 kilometers wide.*

NASA

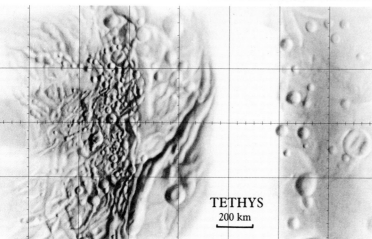

USGS

TETHYS
200 km

Contrary to appearances, Tethys (left) is being lit nearly full-on, and the large central feature (which may be a crater) may be caused more by reflectivity than by relif. It is about 180 kilometers across. Tethys is traveling from left to right and its leading hemisphere has apparently been coated with some dark material.

We're standing on the night side of one of the small satellites that orbit just beyond the outer edge of Saturn's rings. At one time it was thought that a satellite named Janus orbited near here. Although the one we're standing on, and one or two others, are near where Janus was thought to be, the original sighting of Janus was an error . . . the original Janus itself may exist. Hence our designation of this location as "Janus."

The rings, seen edge-on, light the icy landscape like an enormous fluorescent light fixture (right). The sun is far below the horizon. It is late at night on "Janus." The enormous bulk of Saturn is visible at the right, blocking the light of the stars and the faint star-clouds of the Milky Way. Dim light reflected from the rings makes the cloud bands slightly visible. Saturn is still nearly 160,000 kilometers away.

Ron Miller

1979 J2

Here (below) *we are standing on the shadowed floor of a three-kilometer crater on 1979 J2, looking away from Jupiter. We see the satellite Io passing by. Io is surrounded by the cloud of glowing sodium that stretches along its orbit. The sun is below the horizon, but the far crater rim is dimly lit by light from Jupiter, which is above the horizon behind us. In the farther distance, at the left, are Europa with its bright icy surface and Callisto with its darker soil. The visual width of this telephotolike view is 25 degrees.*

After *Voyagers 1* and *2* flew through Jupiter's satellite system in 1979, scientists studied the *Voyager* photos carefully and discovered three new moons. Thirteen moons had been previously charted, and an image of an apparent fourteenth moon had been photographed from Earth in 1975. It was carried in books for a while as satellite "J14," but later searchers failed to recover it; it may have been only a temporarily captured comet.

The moons 1979 J1 and J3 orbit at the edge of Jupiter's ring, are about thirty-five to forty kilometers (twenty-one to twenty-five miles) across, and may be the sources of some dust in the ring.

But 1979 J2 is larger, about seventy kilometers (forty-three miles) across, and is located farther out from Jupiter, between the orbits of Amalthea and Io. The interesting thing about 1979 J2 is that it is the satellite closest to Io, and therefore offers a close look at the changing phenomena of that strange world. When Io is at its nearest approach to 1979 J2, it covers one degree in that moon's sky, and Io's patchy sulfur lava flows are easily visible. When Io is a crescent, flashes mark volcanic explosions on its dark side, and sunlight illuminates umbrellas of ash shooting out in clouds along the edge of Io's disk. At other times, Io's surrounding cloud of sodium atoms emits its characteristic yellow glow.

The surface properties of 1979 J2 are uncertain because it was not photographed at close range by *Voyager 1* or *2*. However, Amalthea, one in from 1979 J2, has an unusual reddish color attributed to atoms and molecules of sulfur and sulfur compounds that were blown off Io, spiraled inward, and were plastered on the neighboring inward satellites. Indeed, 1979 J2 ought to be in an even better position to catch debris from Io, and may have a similar crust of sulfurous material over the older rocky material from which it was originally formed.

Though it is interesting to speculate about visiting 1979 J2 and the other innermost moons of Jupiter, we must remember one of the special perils of Jupiter's system. Jupiter's magnetic field, which extends out to the region of its inner satellites, contains many high-energy atomic particles similar to those trapped in Earth's magnetic field in the Van Allen belts. A much higher concentration of them is trapped in Jupiter's field.

DIONE

Dione is silhouetted against Saturn's russet clouds (above). Dione's trailing hemisphere is dark; wispy streaks make it look rather like a star sapphire.

Large impact craters dominate this view of Dione (above). The bright radiating pattern may consist of ice deposits along fissures.

Twice during Saturn's twenty-nine-year trip around the sun, the sun is lined up with Saturn's ring plane and satellites. During this "season," the satellites can eclipse one another. In this view from the north pole of Dione, we see the next inner satellite, Tethys, passing before the sun.

The glow around the sun is its corona and zodiacal light is scattered off meteoritic dust out to the distance of the asteroid belt. Tethys subtends an angle of about two-thirds of a degree. To the right is the crescent Saturn, a globe subtending 18 degrees.

With the rings backlighted and the solar glare shielded by Tethys (below), we can dimly see the outer extension of the rings, a sheet of dust extending several Saturn radii beyond the main rings. Enceladus is seen in crescent phase in front of Saturn.

DIONE

The largest crater in this Voyager 1 *photograph of Dione* (above) *is less than 100 kilometers across. Sinuous valleys are probably faults in the moon's icy crust.*

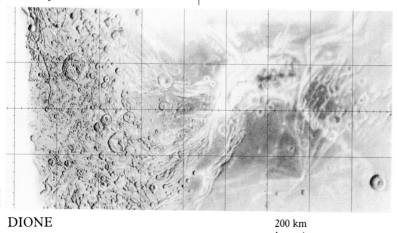

DIONE

200 km

AMALTHEA

A baleful dark eye gazes down on the brick-red surface of Amalthea. It is Jupiter's Great Red Spot, outlined in luminous clouds, only 180,000 kilometers (111,000 miles) away. Electrical storms—of hydrogen bomb intensity—flicker over the giant planet's night side.

Amalthea is a small moon, much closer to Jupiter than any of the larger Galilean satellites. In shape it is rather like a potato, 270 by 170 by 155 kilometers (167 by 105 by 96 miles). Its long axis is lined up toward Jupiter, pulled like a compass needle by the powerful tidal forces, one end always pointing at the planet as it swings in its twelve-hour orbit.

Amalthea is only red on the surface. A little digging might reveal a much lighter color. The scarlet coloring of both the Spot and Amalthea may be a layer of sulfur contaminants blown off Io by meteorites and volcanic eruptions that spiral in toward Jupiter, some of which is intercepted by Amalthea.

The rugged, irregular landscape around us seems to be the shattered remnant of an originally larger body, sculpted by meteorite impacts and coated by debris from Io.

Amalthea

CERES

Contrary to first impressions, "the world Ceres" is more than a dubious pun. It serves to point out that many planetary objects may not be planets according to traditional definitions, but are still large enough to be called worlds. They have all the planetary properties —high gravity, atmospheres in some cases, and extensive surfaces with a variety of geological features.

When *Voyager 1* flew by Jupiter's satellites, they were each revealed to be fascinating and complex. Prior to that time, such small objects had often been dismissed as dead worlds, covered by craters: one crater like another, one moon like another. *Voyager* showed that each of them was different, with its own set of interesting features.

Ceres is a small world, barely larger than our 1,000-kilometer limit. But it is almost twice as large as any other asteroid, and was the first to be discovered. It is located in the middle of the asteroid belt, between Mars and Jupiter. Some scientists believe that when the planets were forming, the asteroids were also involved in a process of accumulation, and that Ceres was

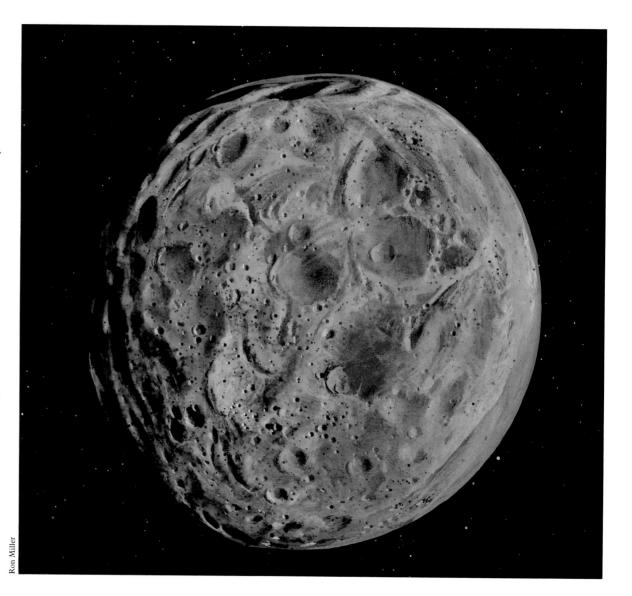

Ron Miller

well on its way to sweeping up the other asteroids and becoming a small planet. If this process had continued, most of the smaller asteroids in the belt would eventually have hit Ceres and been absorbed by it, adding to Ceres's bulk.

Jupiter, however, the most massive planet, was growing nearby at the same time, and its gravity disturbed the orbits of the asteroids, often forcing them into collision courses. These collisions blew more mass away from Ceres than they added to it, preventing Ceres from getting started down the road to planethood. Indeed, many large asteroids were blown apart entirely when they hit other asteroids of comparable size. So, due to Jupiter's interference, the asteroid belt area was left with innumerable small fragments, a couple of dozen bodies bigger then 100 kilometers (62 miles), and only one object bigger than 600 kilometers—the world Ceres.

Ceres and its neighbors in the asteroid belt provide a tableau of what was happening in the solar system 4.5 billion years ago, as planetesimals littered the spaceways. Though the asteroid belt contains thousands of asteroids, they are small enough and far enough apart so that an observer on Ceres sees a sky similar to our own. Ceres's sky is certainly not filled with moonlike, tumbling worlds—the way films often depict "asteroid swarms." Every few months we might notice a bright "star" drifting across the sky from night to night; it would be a neighboring asteroid. And every million years or so, we might get a substantially closer look at an asteroid neighbor.

The surface material of Ceres is very dark, reflecting only about 6 percent of the sunlight that strikes it. It resembles the material in the type of meteorite known as a *carbonaceous chondrite. Chondrite* is a common meteorite containing little glassy *chondrules* believed to have formed as molten droplets caused by the heat of asteroid collisions in space. *Carbonaceous* is a dark subtype of chondrite, blackened by abundant carbon-bearing minerals. Many of these minerals contain *water of hydration*—water not in the form of moisture, but bound up molecule by molecule in the crystal structures of minerals.

Supporting the possibility of such minerals on Ceres, astronomers have spectroscopically identified the presence of water of hydration on Ceres. This in turn supports the idea that water and ice were more abundant in and beyond the central asteroid belt than in the inner solar system. Carbonaceous chondrite meteorites may themselves be fragments knocked off asteroids that formed in this region.

A landing on Ceres might give us an opportunity to test some of these ideas, and to investigate a world that was once on its way to becoming a planet, but was rudely interrupted by Jupiter—a fossil relic of the days of planet formation.

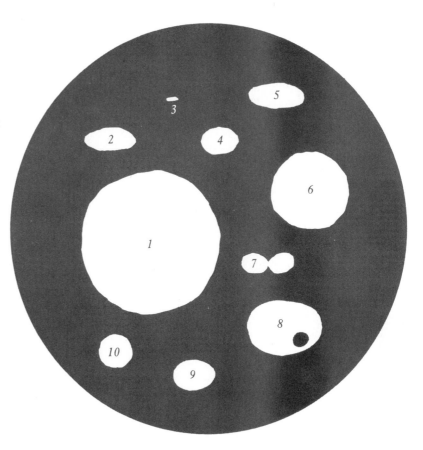

Ten major asteroids compared to Earth's moon (solid color): *1, Ceres; 2, Camilla; 3, Eros; 4, Hygeia; 5, Eunomia; 6, Vesta; 7, Hektor; 8, Pallas and its satellite; 9, Juno; 10, Chiron.*

153

154

CHIRON
& PALLAS

The mysterious asteroid Chiron is seen at left as it passes within 6 million kilometers of Saturn. Its erratic orbit, which takes it as far out as Uranus—farther than any other asteroid—is more like a comet's than an asteroid's. Chiron is somewhere between 100 to 320 kilometers (62 to 198 miles) across, 10 to 100 times the estimated size of comets. Its dark surface may cover an icy core.

Seen from directly over a pole, Saturn resembles a giant bull's-eye. Orange Titan is at the upper left, 1.2 million kilometers (7,440,000 miles) from Saturn.

Pallas (right) is one of the largest asteroids (580 kilometers across), and was the second to be discovered. It was recently found to possess a satellite, which we see here casting an elongated shadow. Before this, it was thought unlikely for asteroids to have moons. Pallas's surface is very dark, but still reflects enough light to dimly illuminate the night side of its tiny moonlet, 90 kilometers in diameter and 300 kilometers from the cratered surface of Pallas. The little satellite is too small to be round.

Ron Miller   W.K. Hartmann

155

ENCELADUS

W.K. Hartmann

Of all the satellites in the solar system, Enceladus, a Saturnian moon, has the best view of a flock of neighboring satellites. If we turn away from Saturn and look in the opposite direction, we will occasionally see conjunctions, or groupings, of Saturn's outer satellites. In the view at right, the sun is behind us. We see the next outer moon, Tethys, subtending an angle of 1.2 degrees, over twice the size of our moon seen from Earth. Beyond Tethys are the moons Dione (*upper right*) and Rhea (*lower left*), each about half the apparent size of our moon. In the distance is massive Titan, with its dense atmosphere of ruddy clouds. On the horizon is the rim of an impact crater that has penetrated into subsurface ice and blown out bright, icy rubble. The visual angle of this "telephoto" view is a narrow 20 degrees.

Turning our backs on the satellites, we find Saturn looming behind us—filling the sky in this 20-degree view (*left*). We're standing at the brink of a crater in Enceladus's icy surface. The night side of Saturn is dimly illuminated by light reflected from the rings. Since we are in the same plane, the broad, brilliant ring system is reduced to a bright, narrow line.

The complex shadow of the rings covers the upper quarter of Saturn's crescent.

Ron Miller

MIMAS

Mimas is the innermost of Saturn's large moons, an icy globe 390 kilometers (241 miles) wide. Although, seen from Earth, it is very faint in the glare of nearby Saturn, it was discovered in 1789 by Sir William Herschel. For years it remained something of an enigma because measurements of its diameter were uncertain and, consequently, calculations of its mean density (mass divided by volume) were unreliable. More recently, data indicated a density much less than that of ice, and writers speculated that Mimas might be a ball of dust, or a chunk of pumicelike material.

The *Voyager* visit to Saturn in 1980 showed that Mimas is a solid body, very heavily cratered, with a mean density only slightly greater than that of pure water ice, implying that Mimas is mostly ice, perhaps with a small amount of rocky material thrown in. The two nearby moons Enceladus and Tethys have similar compositions. Mimas's surface material has not been identified by its spectra because of its faintness, but neighboring satellites have surfaces of frozen water. Taken together, these facts indicate that Mimas is an enormous iceball, inside and out.

Mimas's extremely cratered

A 130-kilometer crater makes Mimas look like a giant eye (right). This feature is more than a quarter of the moon's diameter . . . probably the largest crater diameter-satellite diameter ratio in the solar system. The black spot is a reseau mark made by the camera.

landscape suggests that its surface is very old, having absorbed impacts since the days of planetary formation. Mimas probably isn't large enough to have generated high internal temperatures and, consequently, geologic activity. But one of the impact craters is truly extraordinary: it is about one-third the diameter of Mimas itself. It appears to be moderately fresh, is well-formed, and lies on the leading side of the moon. It's about 130 kilometers (80 miles) across, 9 kilometers (5.6 miles) deep, and has a broad central peak 4 kilometers high (13,000 feet). It is similar in form to but slightly larger than the familiar lunar craters Tycho and Copernicus—both of which are about the size of Yellowstone National Park. It is so large in relation to Mimas that it must have severely damaged the satellite when it formed.

Mimas is of further interest because of its gravitational influence on particles in Saturn's

NASA

rings—an effect called *resonance*. Mimas goes around Saturn in 22.6 hours. Particles in the *central* part of the rings go around Saturn twice in this time. This means that particles in that particular region of the rings encounter Mimas—and its gravitational pull—every other trip around Saturn. The repeated pull due to gravity (resonance) makes the particles move farther and farther from Saturn; random pulls would not have this effect.

Saturn fills Mimas's sky like a golden waterfall (right). Since Mimas always keeps one face permanently locked onto Saturn, the ringed planet never sets. As Saturn goes through its phases, night sweeps over the striped bulk like a theater curtain being lowered. The rings, seen edge on, are a brilliant white, and their shadow splits the awesome planet in two. We're only about half the Earth-moon distance from Saturn, which looms more than seventy times larger than a full moon back home—completely filling this view.

Ron Miller

159

MIMAS

Resonant gravitational pulls can strongly affect orbital motions. In this case, they act to kick particles out of this particular zone of Saturn's rings —the site of Cassini's Division. Indeed, many ring divisions might be explained by resonance. Any body with a large enough mass that is close to the rings (like Mimas) could cause the gaps; perhaps others are due to smaller, closer satellites like S11 that have been discovered at the ring boundaries.

Craters as small as two kilometers are visible in this view of Mimas (above).

The heavily cratered surface of Mimas as photographed by Voyager 1 *(above). A long, narrow trough about five kilometers wide cuts through the rugged landscape.*

160

Cliffs nearly ten kilometers high, among the highest and steepest in the solar system, surround the central mountain of Mimas's giant crater (above and right).

MIMAS

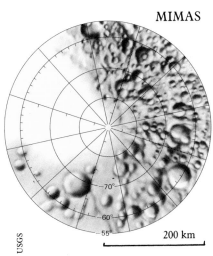

70°

60°

55°

200 km

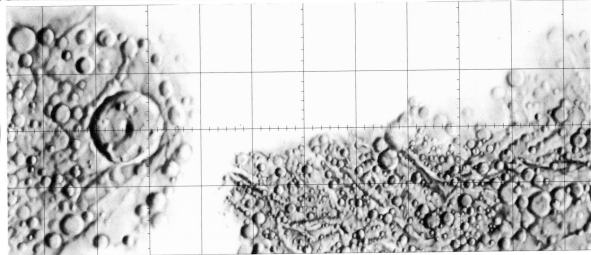

HEKTOR

Although most asteroids are located in the asteroid belt between Mars and Jupiter, some, ranging in size from small to moderate, orbit the sun in other parts of the solar system. Among the most interesting of these mavericks are two groups that follow the same orbit as Jupiter —one 60 degrees ahead of Jupiter and the other 60 degrees behind. The gravitational forces of both the sun and Jupiter conspire to hold them in place.

Since astronomers named these asteroids after heroes in the Trojan wars, they have become known as the Trojan asteroids.

The largest Trojan is Hektor, and Hektor has an unusual property: when viewed from Earth, it varies in brightness by a factor of three. Variations like this are not unprecedented; but in this case some strange conclusions resulted.

When other asteroids vary in brightness (and usually by much smaller amounts) there are usually one or two suggested explanations. First, the asteroid might be elongated, tumbling end over end and presenting first a broad side and then a small end. This theory has proven correct in some cases. These objects—usually much smaller than Hektor—are probably

splinter-shaped fragments, like Eros, broken by collisions between larger "parent" asteroids. Another asteroid might be more or less round, but with one dark and one light hemisphere, much like Saturn's Iapetus.

In the mid 1970s William K. Hartmann and Dale Cruikshank proved that the first theory applies; Hektor is strongly elongated. Since Hektor is the largest Trojan, and its neighbors are no more than half its size and are round, unlike fragments. We suggested that Hektor might have formed in an unusual, low-speed collision where two "normal" Trojans stuck together in a "peanut" shape.

Other studies have suggested that, allowing for gravity, tidal stretching, rotation, and weak, fractured material, Hektor would have distorted into a shape like two eggs stuck end-to-end. Perhaps an initial collision produced a loosely consolidated, fragmented mass, which then deformed into the weird configuration we suspect. Alternatively, Hektor may be *binary*, its two halves slightly separated and orbiting around one another. We may not truly understand the secrets of Hektor's strange shape until

some man-made vehicle approaches it and takes photographs and samples. Perhaps when we do visit this odd little world, it will turn out to be twin-lobed. We'll be able to hike to the contact region, where a mountain of rock 150 kilometers (93 miles) wide will hang directly over our heads, and *leap* (in a gravity less than one percent that of the moon) from one worldlet to another.

Asteroid Hektor (above) is 300 kilometers (186 miles) long; it is highly elongated. It may be a double object consisting of two egg-shaped lobes crushed together or just touching. In this picture, we see two different views of Hektor, about an hour apart in its rotation. The same craters are seen from different perspectives. The stars of the Big Dipper are in the background.

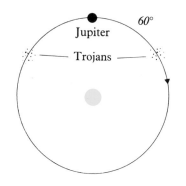

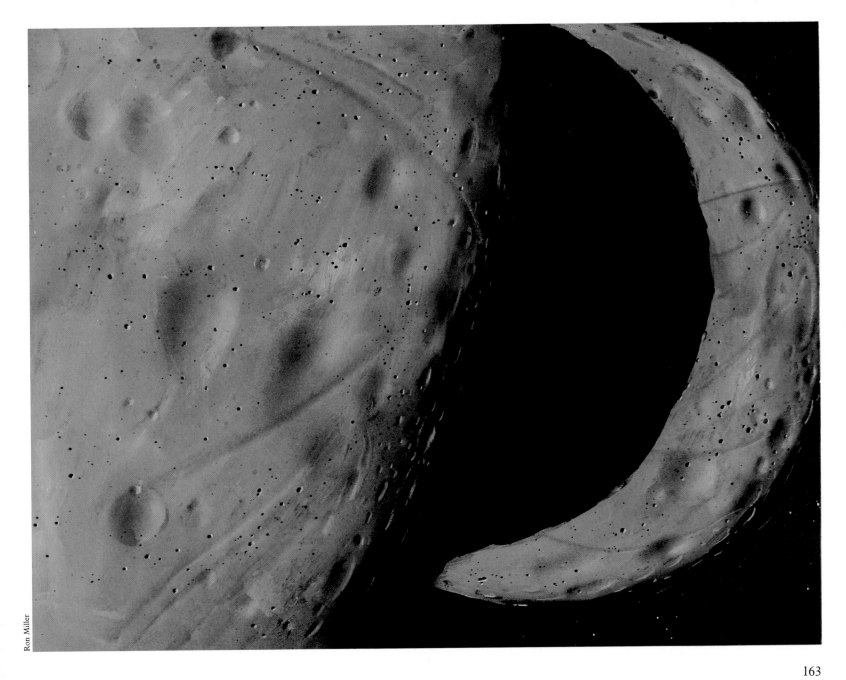

Ron Miller

Ron Miller

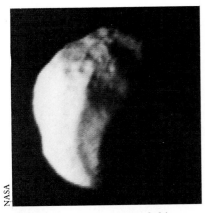

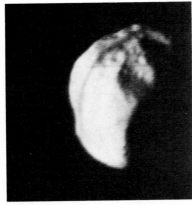

A curiosity of S11 was revealed in a pair of Voyager 1 *photos taken thirteen minutes apart (above). The pictures revealed the shadow of a ring stretching across the cratered surface of S11, moving from one edge of S11 to its central region in that thirteen-minute period as S11 moved relative to the rings and the sun. In the view at left, we are standing on S11. The ring casting the shadow in the* Voyager *pictures was apparently an outer portion of a thin ring just inside S11's orbit, rather than the main, bright portion familiar to us on Earth. Here we look toward this part of the rings from S11, with the sun obscured behind them. They stretch across the sky, dimly backlit by the sun, casting a broad shadow whose darkest portion is only a few hundred meters wide. As they cross the zodiacal light, the rings are silhouetted against a brighter background and appear as a dark line. The darkest phase of the eclipse, unlike anything seen in the Earth-moon system, lasts only a few seconds, and repeats occasionally during S11's sixteen-and-a-half-hour trip around Saturn, but only occurs during the few weeks every fifteen years when the sun aligns with Saturn's ring plane.*

When Saturn's rings are tipped toward Earthbound observers, the glare from the rings and the disk of Saturn make it difficult to see any inner moons. But every fifteen years or so, the rings align with Earth for a few weeks, and telescopic observers have a chance to detect fainter inner moons than they can usually see. In 1980, *Voyager 1* flew past Saturn and firmly established the existence of several of these moons. That gives us several new moons to add to the nine satellites that have been cataloged since 1898.

Of these, S10, S11, S13, S14, and S15 are designations given to small- to moderate-sized moons on the edge of the rings. S12 is the designation of an unusual moon located 60 degrees ahead of Dione and in the same orbit. None of these satellites have received formal names, which must be adopted by the International Astronomical Union. They range from about 40 to 200 kilometers (25 to 124 miles) in diameter.

One of these moons, S11, is located just outside the F ring, the thin "braided" ring which in turn is located just outside the outer edge of the main bright ring system. *Voyager 1* obtained good close-up photos of this moon, revealing it to be an irregular potato-shaped object about 180 by 80 kilometers (112 by 50 miles) in size. Like other small planetary bodies, it has been bombarded by meteorites and is heavily cratered. S11's orbit is only about fifty kilometers from the concentric orbit of the next satellite out, S10. The orbits are closer than the satellite diameters, which means that if they pass one another, they should collide. Yet they exist. Dynamicists believe that in the complex gravity field shared by the rings and other satellites, S10 and S11 go through a peculiar "dance" as they approach each other, eventually exchanging orbits.

S11 is so close to the main ring system that the outer edges of the bright rings, as seen in the sky of S11, stretches about 128 degrees across the sky on the side of S11 that faces Saturn. The F-ring, the faint ring just outside the main rings, covers 135 degrees. Only the widest-angle lens could capture Saturn and its rings as seen from S11.

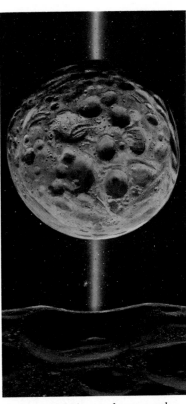

As S11 and S10 overtake one another, we are witness to the unnerving sight of a satellite bearing down on us, (above) apparently on an inevitable collision course.

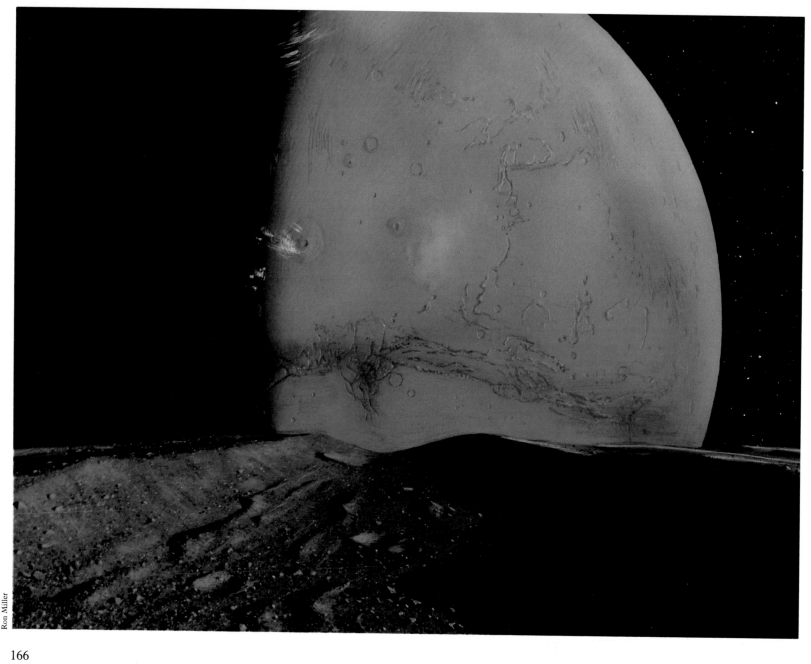

166

PHOBOS

Phobos is a curious, potato-shaped lump of rocky material, twenty-seven by twenty-one by nineteen kilometers. Its surface is a nearly black material that reflects barely 3 to 5 percent of the sun's light that strikes it. It is darker than an asphalt parking lot. It appears to be similar in composition to a carbonaceous chondrite, rich in

Less than 6,000 kilometers (3,720 miles) away, Mars looms in the sky of Phobos, its nearer moon (left). We're standing in one of many shallow grooves that scar Phobos's surface. Millions of years of scouring by micrometeorites have made Phobos, and its sister moon, Deimos, two of the darkest bodies in the solar system. Phobos is only 27 kilometers (16.7 miles) long, too small to have formed into a sphere: it greatly resembles a potato. Mars's terminator, the dividing line between night and day, is sweeping from east to west, unveiling the giant canyon Valles Marineris and bisecting Pavonis Mons, an enormous volcano. Early morning clouds streak and mottle the dawn line and a yellow patch of dust storm stirs in the center of the planet.

The visitor must be very careful on this tiny moon. Its gravity has a tenuous hold on things. A hard leap could leave you literally hanging for several minutes, and a good swing with a bat could put a baseball into orbit.

carbon compounds and chemically bonded water. This material is most commonly found in asteroids in the outermost part of the asteroid belt and in the most distant of Jupiter's moons, which would seem to indicate that Phobos did not form in orbit around Mars, but rather originated as an asteroid captured in a Martian orbit during the closing days of planetary formation.

Phobos is pocked and sculpted by craters. Large chunks of its surface have been blasted away by impacts. The largest crater, Stickney, stretches eight kilometers across—one-third the diameter of Phobos itself. It's the result of an impact nearly large enough to have split Phobos in half. Grooves that stretch in parallel pattern around the little moon, and radiate from Stickney, may be evidence of the fracturing that must have resulted from this impact.

Strangely, the grooves are not uniform, but are pocked by little craters with raised rims. Some grooves are essentially chains of craters. The craters don't seem to be simple collapses (which have flat rims) or impact sites (which happen at random). They look like the sites of minor eruptions.

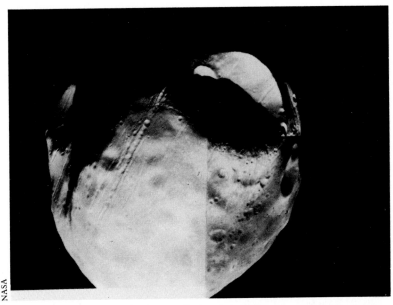

NASA

Perhaps Phobos once had enough water or ice in its interior to generate gas pressure. The Stickney fractures may have been the sites of eruptions that blew away puffs of surface dust in Phobos's weak gravity (where only forty kilometers per hour is fast enough to launch anything permanently off Phobos into space).

Phobos (above) was photographed by Viking 1 from 612 kilometers away. This is the side that permanently faces Mars. The large crater is Stickney. The shadow of Kepler Ridge partially covers Hall Crater. Many grooves are also visible.

DEIMOS

Deimos is the outer of Mars's two moons. Smaller than its sister, Phobos, it is also potato-shaped; its dimensions are 15 by 12 by 11 kilometers (9 by 7.4 by 7 miles). Again like Phobos, it has a dark, nearly black surface, probably consisting of rocks and powder of some carbon-rich minerals, similar to those on a carbonaceous chondrite meteorite. As mentioned in the description of Phobos, this may indicate that Deimos was originally an asteroid, captured by Mars's gravity into an orbit around the red planet.

But these are the only similarities between the sister moons. Although they seem to be made of the same type of rock, their surface textures are very different. Both are cratered, but Phobos has a lumpier appearance and many enormous craters. Deimos has a more muted surface; its craters have low, rounded rims. In the rolling tracts between craters are scattered stadium-sized pits and house-sized boulders.

Why should two neighboring satellites, with nearly identical compositions and virtually no internally-caused geologic activity, have such different-looking surfaces? One reason may be that debris knocked off

NASA

the moons by meteorite impacts goes into orbit around Mars and ends up hitting each satellite again in 100 to 10,000 years. The reaccumulation rates and speeds are somewhat different for each moon and might explain their differing surfaces.

Alternatively, one satellite may have been hit more recently, allowing the other more time to build up a smooth, powdery surface by means of micrometeorite sandblasting.

If you pick up a rock, hold it at eye level, and let it fall on Diemos, it won't plummet to the surface, as it would on Earth. It also won't hang in front of you, as it would on board a free-floating spacecraft. In Deimos's weak gravity, it will settle slowly toward your feet, as though it were being lowered on an invisible thread. Some thirty seconds after dropping it, it will reach the ground—fifty times longer than it would have taken back home.

The bland, micrometeorite-eroded surface of Deimos (left), photographed from fifty kilometers away by Viking 2. The smallest features visible are about eight meters across.

Ruddy Mars-light illuminates the landscape around us (right). We've taken the short hike from the sunlit side of Deimos—scarcely twenty kilometers takes us halfway around the moon—in order to see the dark, rolling surface lit by the Red Planet. Smaller than Mars's inner moon, Phobos, Deimos is a potato-shaped object only fifteen kilometers long. The horizon is only a few hundred meters away. Like Phobos, Deimos is one of the darkest objects in the solar system.

We get a magnificent view of Mars from here. Since one side of Deimos always faces Mars, we can watch Mars go through all its phases every thirty hours. We can see the dark, triangular area at the left that is called Syrtis Major, one of the first Martian features to have been seen from Earth. A haze of water vapor hoods the north polar cap, while clouds streak the surface. Wave clouds are forming over a group of volcanoes near the terminator at the top. Like Earth, Mars presents an ever-changing face. Clouds come and go, the polar caps shrink and expand, the dark areas mysteriously change size and shape, and an occasional dust storm veils the entire planet in an opaque yellow haze.

Ron Miller

AMOR & APOLLO

Most asteroids are located in the asteroid belt between Mars and Jupiter, but others travel in regions well outside the belt, and still others move at least partially in the inner solar system, inside the belt region. Members of a subgroup of these, *Amor asteroids*, follow orbits that bring them closer to the sun than the planet Mars—that is, their orbits actually cross over Mars's orbit. At their farthest from the sun, most of them follow paths that intersect the belt.

Amor asteroids are named after one of the prominent early members of their group to be recognized, the asteroid Amor, discovered in 1932. Several dozen Amors, or Mars-crossers, as they are sometimes called, are known. Among the largest are Ganymed —not to be confused with Jupiter's satellite Ganymede— (twenty-five to thirty kilometers in diameter) and the splinter-shaped Eros (measured at seven by nineteen by thirty kilometers), which may be a fragment from a large-scale collision. Other Mars-crossers are typically five to ten kilometers across, but future surveys will surely turn up even smaller ones.

The fact that Amor asteroids cross the orbit of Mars doesn't necessarily mean that they all come very close to it, or are likely to hit it in the next million years or so. At the points where they cross Mars's orbit, moving toward or away from the sun, many of them are far above or below the plane of the solar system, and thus don't approach Mars. Some, however, do. In any case, gravitational disturbances by the planets are likely to change Amor orbits slowly over millions of years, so that eventually they will move into configurations in which close approaches and impacts *are* possible. Those that don't directly hit Mars may be thrown into new courses by close encounters with Mars; these new courses may take them on Earth-crossing orbits, or orbits that allow close approaches to some other planet. Therefore, Amor or other asteroids that venture into the inner solar system are likely to end their days within 10 or 100 million years by colliding with one of the planets, creating large craters on the surfaces of Mars, Earth, the moon, or on some other world.

Apollo asteroids, related to Amor asteroids, differ in that they drop far enough into the inner solar system to cross over the orbit of Earth. Their closest approach to the sun (typically 0.5 to 0.9 astronomical units) is inside Earth's orbit, and their

W.K. Hartmann

farthest from the sun (typically 2 to 4 AUs) is usually in the asteroid belt. Apollo asteroids, like Amors, are named after one prominent example of their group. They are sometimes called Earth-crossers, or, more ominously, Earth-grazers; just as some Amor Mars-crossers come close to Mars, some Apollo Earth-crossers can come very close to Earth.

Apollo asteroids tie in with another phenomenon: meteorites. Apollo asteroids are, essentially, giant meteoroides. Of the several dozen Apollo asteroids now known, the biggest are about eight to ten kilometers across, and nearly all are bigger than one

Here (left) *we are standing on a crater rim of a one-kilometer Apollo asteroid that is passing by the Earth-moon system (upper right). The sun has just set below the rim, and the pearly corona and zodiacal light glow prominently. This asteroid happens to be crossing Earth's orbit at a time when a major comet is passing through the inner solar system. The comet is seen directly above the sunset, with its gas and dust tails pointing away from the sun. Earth and the moon are silhouetted against the comet's tail.*

The asteroid we're on is one of a newly discovered class; these asteroids have minor satellites of their own. The satellite is above the horizon to the right. Venus is prominent in the coronal glow of the sun, part of Orion is visible in the lower sky, and the Hyades star cluster is in the upper center sky left of Earth.

Ron Miller

Venus is seen in the far distance from the asteroid Apollo (above), *one of the rare asteroids that crosses the orbit of that planet. Apollo is influenced by the gravitational fields of both Earth and Venus, and will probably eventually, fall onto one or the other of the two planets. Apollo's elliptical orbit will take it beyond the orbit of Mars before it begins its swing back toward the sun.*

kilometer. Many smaller ones also exist, all too tiny to detect with telescopes, but they are so numerous that many hit Earth each year, breaking into pieces in the atmosphere and hitting the ground as meteorites.

Although spectroscopic studies of Apollo asteroids reveal many different rock types among them, all these rock types are similar to various meteorites, and some of them almost perfectly match the spectroscopic properties of certain common meteorite types. This means that some of the

meteorites in our museums are probably fragments of Apollo asteroids—or at least fragments of the same parent body that the Apollo asteroids are themselves part of.

Some of the Apollo asteroids are probably fragments of asteroids from the asteroid belt, perturbed into Earth-crossing orbits by gravitational disturbances from Jupiter or Mars, the planets on either side of the belt. Other Apollo asteroids may have a quite different origin. They may be the burnt-out, rocky remnants of comets. Fresh comets are believed to be mostly ice and to contain a certain proportion of rocky material. But often, comets get perturbed by planets into orbits in the inner solar system where the ices eventually evaporate or sublime into space, leaving the rocky material in a loosely consolidated clump. Some Apollos may be such clumps.

Apollo asteroids occasionally crash into Earth, causing monstrous explosions like the atom bomb-sized one in Siberia in 1908. That blast flattened trees in an 18-mile circle, knocked a man off a porch 38 miles away, and was heard more than 600 miles away. Statistics suggest that such impacts may occur every century or so—but they usually take place in the ocean, and leave little or no historical record.

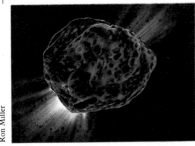

Ron Miller

The tiny, kilometer-and-a-half-thick chunk of rock called Icarus (above) *is one of the only bodies in the solar system that regularly passes inside the orbit of Mercury five times closer to the sun than Earth is, twice as close as Mercury. Icarus's sunlit side is receiving over twenty-five times the heat and radiation Earth does. But it will have plenty of time to cool off, since Icarus' highly elliptical orbit also takes it beyond Mars.*

AMOR & APOLLO

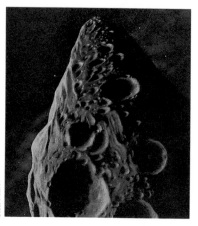

Boxcar-shaped Eros (right) forever tumbles end over end. It is probably a thirty-kilometer-long splinter shattered off a larger body by an ancient collision. It is one of the Amor group of asteroids that are sometimes called the Mars-crossers.

Ron Miller

W.K. Hartmann

We are standing (left) on the surface of a five-kilometer Amor asteroid that is making a close approach to Mars at a distance of some 55,000 kilometers (34,100 miles). We have an unusual view of Mars, because the asteroid happens to be passing "under" the south pole. Mars's south polar ice cap shows up as a white ice field at the center of the Martian disk, half illuminated by the sun. In the background is the northern constellation of the Big Dipper, which seems to have two new, bright stars. Above the handle (upper left), nearly 7 Martian radii out from Mars, is Mars's satellite Deimos, shining about as bright as the brightest star, Sirius. To the lower right of Mars, at about 2.7 radii, is Phobos, still brighter. The surface of our asteroid is composed of dark, rocky material and the pulverized dust of many impacts. Two eroded craters are in front of us, and a fresher, larger crater is seen on the nearby horizon. Gravity is so weak that if you picked up a rock and threw it hard, you could put it into orbit around the asteroid. But it would be wise to duck before it returns.

COMETS

Comets are icy planetesimals, or planetary building blocks, that formed in the cold, outer part of the solar system while the planets were growing, 4.5 billion years ago. Most of them are believed to have diameters in the range of 1 to 50 kilometers (.6 to 31 miles) and to be composed of a mixture of ice and rocky material. This composition has been deduced from spectroscopic studies of the gases and dust that form a comet's distinctive tail. Some years ago, the Harvard comet expert, Fred Whipple, coined the term "dirty iceberg" to describe a comet. Many analysts believe the dirt is mostly carbonaceous chondrite meteoritic material, the type of soil typically formed in the outer solar system.

Dynamical studies as long ago as the 1950s showed that when these objects formed among the giant planets, they occasionally underwent close encounters with Jupiter, Uranus, or one of the other giants. Sometimes these would result in collisions and the icy planetesimal would fall into the deep atmosphere of the planet, disappearing in a fleeting blaze. More often, it would get flung into the outermost solar system, thousands of astronomical units from the sun,

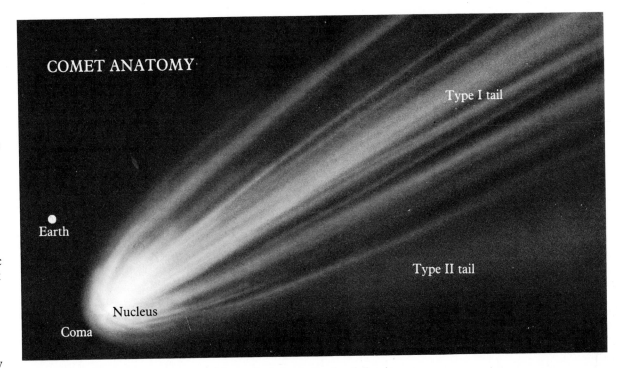

COMET ANATOMY

Type I tail

Type II tail

Earth

Nucleus

Coma

just as *Pioneer* and *Voyager* space vehicles got kicked outward as they passed by Jupiter. The result was that many icy planetesimals were hurled into deep freeze during the early days of the solar system, where they were too far from the sun for their ices to be heated or evaporated by sunlight. But their orbital motions, sometimes disturbed by the nearer stars, inevitably led some of them to move back toward the sun for a trip through the solar system.

If the comet travels past the asteroid belt into the inner solar system, a glorious thing happens. The abundant frozen water is heated to the point that it begins to sublime rapidly into space. At two to three astronomical units from the sun, the comet begins to develop a thin cloud of gas and dust around itself, and may become brightly visible in the skies of Earth. The solid body itself is called the *nucleus*, and is generally too small to be seen as a disk in even large telescopes. But the cloud of gas around the nucleus, called the *coma*, may become larger than any planet, stretching hundreds of thousands of kilometers across.

The radiation and outrushing gas from the sun push the gas and dust of the comet away from

Caught in a celestial searchlight (overleaf), Earth and the moon pass through the tail of a comet. The glittering head is thousands of kilometers away, and the tail continues on behind us for many more millions of kilometers. Our home planet and its satellite are in no danger. As one astronomer put it, comets are as close as you can get to nothing while still having something. For all of its enormous size —the end of the tail may extend as far back as Mars's orbit—the nucleus is an icy mass scarcely larger than a small asteroid. The tail itself, despite its brilliant fluorescence, has so little gas and dust in it that any cubic meter of it is a more complete vacuum than we can make in a laboratory. The entire tail could be packed into a suitcase.

Comet, Earth and Moon

COMETS

the sun, forming a tail that may stretch millions of kilometers. The tails are often seen to have two components. The brilliant Type I tail is composed of ionized gas and streams away from the sun in straight filaments. The darker Type II tail is composed of dust particles, and is usually broader, more diffuse, and gently curved.

If the comet happens to pass close to a planet, its orbit may be drastically changed once again. It might be thrown clear out of the solar system into interstellar space, never to return. Or, it might be slowed down into an orbit that stays in the inner or middle solar system among the planets. Once on such an orbit, it might pass close to the sun every ten, twenty, or eighty years instead of once every few million years. Such comets are called *short period comets*. Halley's comet is the most famous of these. It drops into the inner solar system every seventy-five years or so, and has been observed doing so for more than a thousand years.

Short period comets go through an interesting evolution. Each time around the sun, near *perihelion* (the point in an orbit that is nearest to the sun), they lose considerable amounts of ice,

W. K. Hartmann

We're on the violent surface of an active comet as it enters the inner solar system (left). *The landscape is dominated by rough snowfields of frozen water. The whole surface is eroding away around us—at a rate as high as several inches or even feet per day—as the sunlight sublimes the ice directly into gas. Rocks protecting underlying ice form squat pedestals. Once the rocks fall off, the pedestals quickly erode into slim spires. Fissures split the surface around us, exposing pockets of frozen methane: less stable than water ice, it vaporizes explosively. Jets of gas and debris fountain out of the fissures like miniature volcanoes. What was once a quiet, gray world that seemed and felt as solid as any other has become totally transformed—we can feel the ground itself crawling and shrinking beneath our feet.*

which is "burned away." Some of the rocky dust buried in the surface layers of the comet is also carried away, but larger clumps of soil or rocks (if they exist in any given comet) would tend not to be lost as easily as the gas. After about a thousand trips around the sun (according to some estimates), most of the ice may have been burned away. The comet is now a loosely consolidated rocky core, perhaps emitting an occasional puff when some bit of ice is exposed by a meteorite impact, a fracture, or by a change in rotational orientation. In this way, some short period comets may have been converted into the rocky objects now cataloged as Apollo asteroids.

A trip on a comet would be an extraordinary ride. If we landed on the comet on its way in toward the sun, we would find a dark landscape covered by a crust of blackish, carbonaceous powder and rubble—the rocky material that fell back or remained on the surface as the last bits of exposed ice vaporized on its last trip. As we approached the sun, passing the region of the asteroids and Mars, this surface would get warm enough to sublime the ice below it, and patches of the surface would erupt in outbursts as the gas breaks loose.

Hard patches of consolidated crust or rock would protect the ice beneath, and as surrounding ice sublimes, they would be left standing. A landscape of fantastic

grotesqueries would result, with spires of ice and soil rapidly wearing away.

As veins of ice sublimed away, deep fissures would develop across the comet's surface, and huge sections would break away and become launched slowly outward. Elsewhere, pockets containing greater than normal concentrations of unstable ammonia or methane ice might become exposed and erupt in great geyserlike jets.

Streamers of material would lace the sky, for a distance of thousands of kilometers. A view on the night side would reveal a sky full of glowing streamers converging toward the distant tail. (If the comet got close enough to the sun, the surface fractures might widen and the whole object on which we would be standing could break into two or three subnuclei, surrounded by a swarm of smaller fragments.) If we were shielded from the debris, we might choose to stay aboard and ride the comet back outward into the middle or outermost solar system, watching the activity die down again until all the ices becomes cold and a new crust of soil forms on the surface.

COMETS

This comet nucleus (below) has broken into two main pieces and a cloud of smaller fragments. The break has exposed new ice surfaces to sunlight, causing furious jetting as newly fractured, unstable ice sublimes into gas and sprays outward through fissures in the flying "dirty iceberg." The slower jets are caught in the solar wind that rushes away from the sun (out of the picture to the left). These streamers of material curve around to the right to join in the comet tail. In the far distant background are Earth and the moon.

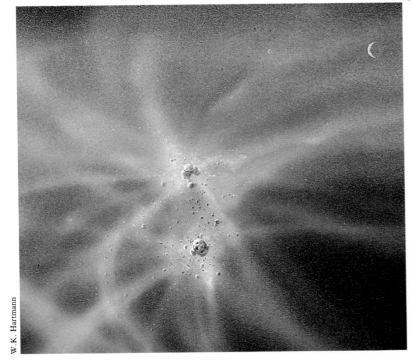

W.K. Hartmann

We are riding a comet that has fallen into the inner solar system (overleaf). When comets leave their haunts in the sunless void beyond Pluto, they loop around the sun and are flung outward —usually never to return. The point of closest approach is called the perihelion. Some comets never come closer to the sun than Earth does. Many venture well within the Earth's orbit. A few have their perihelion distance very close to the sun. They may pass within one or two million kilometers of its fiery surface, far inside the orbit of Mercury. These are the sun-grazing comets, like the one we are on now. In theory, a comet could have a perihelion within the sun's radius; it would soar into the sun and be consumed.

The surface of a sun-grazer gets intensely heated by the sun during its perihelion passage. The time spent close to the sun is so brief that this heat may not have time to "soak" into the interior. The ices on the comet's surface will furiously sublime into space, leaving only the rocky dregs that we're standing on now. After a few trips around the sun—possibly over thousands of years—only a barren hulk will be left, resembling an asteroid. Alternatively, the rapid vaporization of the ice may remove the "glue" holding the comet's rocky material together and it may simply fall apart. Several sun-grazing comets have been seen to go to pieces at perihelion passage. Sometimes these pieces themselves vaporize rapidly and are never seen again.

Our ride will be a hair-raising experience as we watch the sun loom larger and hotter each day, and as our comet's structural integrity weakens. Here we are sheltered from the sun's searing heat by the body of the comet during the time of perihelion passage. The sun itself is hidden below, but an enormous prominence of hot solar gas, glowing with the red light of hydrogen, writhes above the rocky horizon.

177

Ron Miller

Sun-Grazing Comet

A dead comet's spectacular immolation (right): *a fireworks display of meteors. As the ice sublimes, it carrys away particles of soil. The process may eventually make some comets break up into myriads of tiny bits. Most are smaller than a grain of sand. Earth passes through several swarms of meteors every year, and their orbits are associated with the orbits of active or extinct comets. During the most spectacular of these annual meteor showers, hundreds of meteors can be seen every minute.*

GLOSSARY

ALBEDO
Reflectivity or brightness of an object

APOGEE
Point in an orbit farthest from Earth (*apolune:* farthest from the moon; *aphelion:* farthest from the sun; *apapsis:* farthest from any object, general term). See PERIGEE

CONJUNCTION
When two or more planets appear close together in the sky

CRUST
Outermost solid layer of a planet

DENSITY
The proportion of mass to volume: a pound of lead is more dense than a pound of air

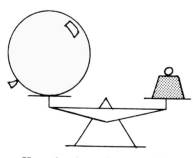

Knowing how *dense* an object is can give you a good idea of what it is made of. Different substances have different densities. Gases have very low densities, liquids higher

densities, and solids like rocks and metals usually have even higher densities. Water is used as the standard for all densities and is given a value of 1. Saturn has a very low density, less than 1, so that we know that it is made of something very light. In fact, it is made up mostly of gases. Earth has a density of 5.5, indicating that it is made of some very heavy substances. If an object has a density of 3, we know that is more than water (1) but less than rock (say, 5). A good guess

would be that it is a *mixture* of rock and water or ice. Adding rock to ice would raise the water's average density, while the ice would lower the rock's, giving us a density halfway between the two.

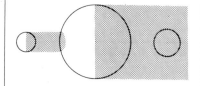

ECLIPSE
When one body passes into the shadow of another

ECLIPTIC
Plane of Earth's orbit; approximate plane of the solar system

GALILEAN SATELLITES
Jupiter's four largest moons (Io, Europa, Ganymede, and Callisto); named as a group for Galileo Galilei, who discovered them in 1610

LATIN PLACE NAMES

To avoid international confusion, or favoritism of any one language, "neutral" Latin is used to describe features on the planets and satellites, just as Latin names are used in biology and botany. This has been standardized so that "mountain" on one map is "mountain" on another. Following is a list of these descriptive terms:

DORSA:	Scarps
MARIA:	"Seas" (singular: MARE)
MONTE:	Mountains (singular: MONS)
PATERA:	Shallow, dish-shaped depression
PLANITIA:	Plains or basins
RUPES:	Ridges
RILLE:	Narrow, linear valley
VALLES:	Valleys

LIMB
Apparent edge of an object

LITHOSPHERE
Solid rocky layer in partially molten planet

MANTLE
A region of intermediate density around the core of a planet

MASS
The amount of material in an object

METEOR
A "shooting star," the streak of light made by an object burning up as it enters atmosphere

METEORITE
A meteoroid when it is on the ground

METEOROID
Any small rocky, metallic, or carbonaceous object in space, usually smaller than the size of a pea

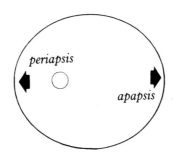

METRIC MEASUREMENTS

meters x 3.281 = feet
kilometers x 0.621 = miles
degrees Celsius x 9/5
    (now add 32) = degrees
                    Fahrenheit
feet x 0.305 = meters
miles x 1.609 = meters
degrees Fahrenheit
  (now subtract 32)
          x 5/9 = degrees
                    Celsius

MILLIBAR
A unit of atmospheric pressure;
the average pressure at Earth's
surface is 1,000 millibars.

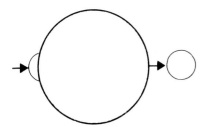

OCCULTATION
When a smaller body passes
behind a larger one

ORGANIC MOLECULE
One based on carbon, usually
large and complex

PERIGEE
Point in an orbit nearest Earth
(*perilune:* nearest the moon;
perihelion: nearest the sun;
periapsis: nearest any object,
general term)

PHASES

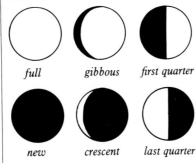

full   gibbous   first quarter

new   crescent   last quarter

PLANE OF THE SOLAR SYSTEM
The plane of Earth's orbit around
the sun. With few exceptions, all
the other planets orbit in or near
this same plane

PLANETESIMAL
Another (and more accurate)
name for an asteroid; also often
used is *minor planet*

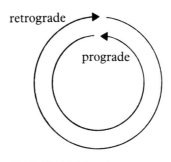

RETROGRADE and PROGRADE
When describing the rotation of a
planet or its movement in its
orbit, retrograde means clockwise
(east to west), when seen from
above the planet's north pole, and
prograde means counterclockwise
(west to east), when seen from
above the north pole

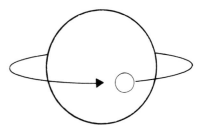

REVOLUTION and

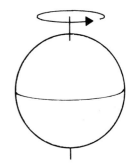

ROTATION
Revolution describes the
movement of one body around
another; rotation describes the
movement of an object around its
own axis

181

GLOSSARY

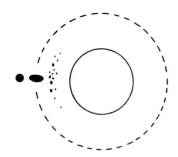

ROCHE'S LIMIT
Distance from a planet within which tidal forces would disrupt a large enough satellite

SATELLITE
Any small body orbiting a larger one

SUBLIME
When a substance, like ice, goes directly from a solid to a gaseous state without first becoming a liquid; dry ice (frozen carbon dioxide) does this on Earth

SUNSPOT
A magnetic disturbance on the sun; it is cooler than the surrounding area and, consequently, appears darker

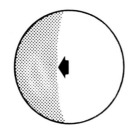

TERMINATOR
Dawn or dusk line separating night from day

TERRESTRIAL PLANETS
Planets resembling Earth in that they are primarily made of rocky material

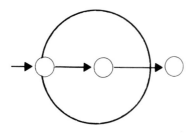

TRANSIT
When a smaller body passes in front of a larger one

VISUAL ANGLES

In order to help our readers visualize what they would see were they to actually visit any of the places described in this book, we have included *visual angles* in the captions of many of the paintings. The horizon in front of you, if you are outdoors, covers 180 degrees from your right to your left. Normal vision, as you stare straight ahead, covers about 90 or 100 degrees of this area, although you are visually aware of—and mainly using—only about 40 degrees of that. (Forty degrees is the angle covered by a picture taken by an ordinary snapshot camera.)

A few of the scenes in the book cover a larger width, 90 degrees or even more. These are *wide-angle* views, the kind that you could take with a wide-angle camera lens. This lens has the effect of making any object appear relatively *smaller* that it would in a normal-angle picture. Our moon, for example, covers half a degree in the sky. It follows that it would thus be twice as large in a 40 degree-wide picture as it would be in an 80 degree-wide picture. Since they have the effect of shrinking distant objects, wide-angle views are only used when they are necessary to show some feature in the landscape or foreground that is too large or extensive to see completely otherwise. In the book, if a view deviates substantially from the everyday snapshot angle, or a visual angle of 30 to 45 degrees, we have pointed this out.

ZODIACAL LIGHT
Faint glow caused by dust in the ecliptic

VITAL STATISTICS

OBJECT	DISCOVERY	ORBITAL PERIOD (r = retrograde)	DISTANCE From Sun (au) or Planet (1000's km)	DIAMETER (km)	SURFACE GRAVITY* (Earth=1)	SURFACE MATERIAL	ATMOSPHERE
MERCURY	Prehistory	86.0 d	0.39 AU	4,880	.39	Basaltic dust & rock	None
VENUS	Prehistory	225.0 d	0.72 AU	12,100	.91	Basaltic & granitic rock	Thick, CO_2, H_2SO_4
EARTH	~ 500 BC	1 yr	1.00 AU	12,756	1.00	Water, Granitic soil	N_2, O_2, H_2O
Moon	Prehistory	27.3 d	384	3,476	.16	Basaltic dust & rock	None
MARS	Prehistory	1.88 yr	1.52 AU	6,794	.38	Basaltic dust & rock	Thin, CO_2
Phobos	1877	7.7 h	9.4	27x21x19	.0009	Carbonaceous rock	None
Deimos	1877	30.3 h	23.5	15x12x11	.0004	Carbonaceous rock	None
CERES	1801	4.6 yr	2.77 AU	1,025	.04	Carbonaceous rock	None
PALLAS	1802	3.63 yr	2.77 AU	583	.02	Meteoric rock	None
JUNO	1804	4.61 yr	2.67 AU	249	.01	Rock & iron	None
VESTA	1807	5.59 yr	2.36 AU	555	.02	Basaltic, meteoritic rock	None
JUPITER	Prehistory	11.9 yr	5.2 AU	143,200	2.6	Liquid H_2 (?)	H_2, He, NH_3, CH_4
1979 J3	1980	7.1 h	126	40	.002	Rock (?)	None
1979 J1	1979	7.1 h	128	35	.001	Rock (?)	None
Amalthea	1892	12.0 h	182	270x170x155	.009	Sulfur layer over rock (?)	None
1979 J2	1980	16.2 h	223	75	.003	Rock (?)	None
Io	1610	1.8 d	422	3,640	.18	Sulfur compounds	Very thin SO_2, S, Na
Europa	1610	3.6 d	671	3,130	.14	H_2O ice	None
Ganymede	1610	7.2 d	1,071	5,280	.15	H_2O ice & dust	None
Callisto	1610	16.8 d	1,884	4,840	.12	Rocky dust & some ice	None
Leda	1974	239.0 d	11,094	10	.0003	Carbonaceous rock (?)	None
Himalia	1904	251.0 d	11,487	170	.004	Carbonaceous rock (?)	None

*Multiplying your weight by this number will tell you what you would weigh on a particular world.

VITAL STATISTICS

OBJECT	DISCOVERY	ORBITAL PERIOD (r = retrograde)	DISTANCE From Sun (au) or Planet (1000's km)	DIAMETER (km)	SURFACE GRAVITY* (Earth=1)	SURFACE MATERIAL	ATMOSPHERE
Elara	1904	260.0 d	11,747	80	.002	Carbonaceous rock (?)	None
Lysithea	1938	264.0 d	11,861	25	.0006	Carbonaceous rock (?)	None
Ananke	1951	r1.72 yr	21,250	20	.0005	Cabonaceous rock (?)	None
Carme	1938	r1.89 yr	22,540	30	.0008	Carbonaceous rock (?)	None
Pasiphae	1908	r2.02 yr	23,510	35	.0009	Carbonaceous rock (?)	None
Sinope	1914	r2.07 yr	23,670	30	.0008	Carbonaceous rock (?)	None
SATURN	Prehistory	29.5 yr	9.5 AU	120,000	1.1	Liquid H_2 (?)	H_2, He, NH_3, CH_4
S 15	1980	14.3 h	136	40x20x20	.0008	(?)	None
S 14	1980	14.6 h	138	220	.006	(?)	None
S 13	1980	15.0 h	141	200	.005	(?)	None
S 11	1980	16.7 h	151	180x80	.004	(?)	None
S 10	1980	16.7 h	151	200x180x150	.005	(?)	None
Mimas	1789	22.6 h	186	390	.007	Mostly H_2O ice	None
Enceladus	1789	1.4 d	238	500	.008	Mostly H_2O ice	None
Tethys	1684	1.9 d	295	1,050	.015	Mostly H_2O ice	None
**S 16	1981	1.9 d	295	20(?)	.0003	(?)	None
**S 17	1981	1.9 d	295	30(?)	.0004	(?)	None
Dione	1684	2.7 d	377	1,120	.022	Mostly H_2O ice	None
**Dione B	1980	2.7 d	377	160	.004	Mostly H_2O ice	None
Rhea	1672	4.5 d	527	1,530	.028	Mostly H_2O ice	None
Titan	1655	15.9 d	1,222	5,140	.14	Ices; liquid NH_3 & CH_4	Thick N_2, CH_4
Hyperion	1848	21.3 d	1,484	290	.006	Ices (?)	None
Iapetus	1671	79.3 d	3,562	1,440	.02	Ice & rock	None

**S 16 and S 17 are both in the same orbit as Tethys; S 16 is about 60° ahead of Tethys and S 17 about 60° behind. Dione B is about 60° ahead of Dione and in the same orbit .

OBJECT	DISCOVERY	ORBITAL PERIOD (r = retrograde)	DISTANCE From Sun (au) or Planet (1000's km)	DIAMETER (km)	SURFACE GRAVITY* (Earth=1)	SURFACE MATERIAL	ATMOSPHERE
Phoebe	1898	r550.0 d	12,960	240	.005	Carbonaceous soil	None
CHIRON	1977	51.0 yr	13.7 AU	350	.005	Carbonaceous soil over ice(?)	None
URANUS	1781	84.0 yr	19.16 AU	51,800	.88	(?)	H_2, He, CH_4
Miranda	1948	1.4 d	130	300	.004	H_2O ice, rock	None
Ariel	1851	2.5 d	191	800	.01	H_2O ice, rock	None
Umbriel	1851	4.1 d	266	550	.008	H_2O ice, rock	None
Titania	1787	8.7 d	436	1,000	.02	H_2O ice, rock	None
Oberon	1787	13.5 d	583	900	.01	H_2O ice, rock	None
NEPTUNE	1846	164.0 yr	30.0 AU	49,500	1.14	(?)	H_2, He, CH_4
Triton	1846	r5.9 d	356	4,000	.06	CH_4 ice	Thin CH_4 (?)
Nereid	1949	360.0 d	5,567	300	.004	CH_4 ice (?)	None
PLUTO	1930	247.0 yr	39.4 AU	3,100	.05	CH_4 ice	Very thin CH_4 (?)
Charon	1978	6.4 d	19	1,300	.02	CH_4 ice (?)	None

CH_4	Methane	N_2	Nitrogen
CO_2	Carbon dioxide	Na	Sodium
He	Helium	NH_3	Ammonia
H_2	Hydrogen	O_2	Oxygen
H_2O	Water	S	Sulfur
H_2SO_4	Sulfuric acid	SO_2	Sulfur dioxide

FURTHER READING

Our Universe, Roy Gallant,
National Geographic Society,
Washington, D.C., 1980

Space Art, Ron Miller, O'Quinn
Studios, New York, 1979

Solar System, Ludek Pesek and
Peter Ryan, Viking, New
York, 1980

Astronomy: The Cosmic Journey,
W.K. Hartmann, Wadsworth,
Belmont, Calif., 1978

NASA Publications covering the
results and photos of various
planetary missions are available
from NASA Headquarters,
Washington, D.C.

The New Solar System
ed. J. Beatty, B. O'Leary,
and A. Chaiken
Sky Publishing Co.,
Cambridge, Mass, 1981.

Moons and Planets,
2nd Edition, W.K.
Hartmann; Wadsworth,
Belmont, Ca., 1982, in press.

Orbiting the Sun, Fred L.
Whipple, Harvard
University Press, Cambridge
Mass, 1981.

Worlds of Rock and Ice,
Clark R. Chapman, Scribners,
New York, 1982 in press.

INDEX

INDEX

INDEX

INDEX

INDEX

INDEX